W0079981

Weather & Climate Services for the Energy Industry

Alberto Troccoli
Editor

Weather & Climate Services for the Energy Industry

palgrave
macmillan

Editor
Alberto Troccoli
World Energy & Meteorology Council
c/o University of East Anglia
Norwich, UK

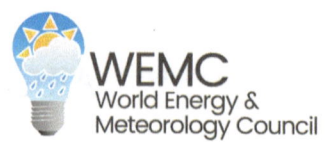

ISBN 978-3-319-68417-8   ISBN 978-3-319-68418-5   (eBook)
https://doi.org/10.1007/978-3-319-68418-5

Library of Congress Control Number: 2017954970

© The Editor(s) (if applicable) and The Author(s) 2018 This book is an open access publication

Open Access This book is licensed under the terms of the Creative Commons Attribution 4.0 International License (http://creativecommons.org/licenses/by/4.0/), which permits use, sharing, adaptation, distribution and reproduction in any medium or format, as long as you give appropriate credit to the original author(s) and the source, provide a link to the Creative Commons license and indicate if changes were made.

The images or other third party material in this book are included in the book's Creative Commons license, unless indicated otherwise in a credit line to the material. If material is not included in the book's Creative Commons license and your intended use is not permitted by statutory regulation or exceeds the permitted use, you will need to obtain permission directly from the copyright holder.

The use of general descriptive names, registered names, trademarks, service marks, etc. in this publication does not imply, even in the absence of a specific statement, that such names are exempt from the relevant protective laws and regulations and therefore free for general use.

The publisher, the authors and the editors are safe to assume that the advice and information in this book are believed to be true and accurate at the date of publication. Neither the publisher nor the authors or the editors give a warranty, express or implied, with respect to the material contained herein or for any errors or omissions that may have been made. The publisher remains neutral with regard to jurisdictional claims in published maps and institutional affiliations.

Cover illustration: © Andrew Taylor/Flickr

Printed on acid-free paper

This Palgrave Macmillan imprint is published by Springer Nature
The registered company is Springer International Publishing AG
The registered company address is: Gewerbestrasse 11, 6330 Cham, Switzerland

"To my wife Elena and my three children, for their sustained lovingly support and understanding."

CONTENTS

NOTES ON CONTRIBUTORS

Mohammed Sadeck Boulahya has more than 35 years of experience managing public regional institutions, mobilising resources, networking and building national capacity for weather and climate services in support of more resilient economies in Africa and the Mediterranean Region. Since 2005, under a number of consultancies, Boulahya has advised the African Development Bank (AfDB) on the conception of a Strategy in Climate Risk Management and Adaptation, organised the First Climate for Development Conference in Addis Ababa, and was instrumental in facilitating the negotiation between African Union Commission (AUC), United Nations Economic Commission for Africa (UNECA) and AfDB for the conception of, and resource mobilisation for, the ClimDev-Africa Programme, leading to its official launch in 2010 during UNECA/ADF(African Development Forum)-VII. Boulahya also co-founded and contributed to the programme implementation as the First Director General of the African Centre of Meteorological Applications for Development (ACMAD) for 12 years.

David Brayshaw PhD, is an associate professor in Climate Science and Energy Meteorology at the Department of Meteorology at the University of Reading and a Principal Investigator (PI) with the UK's National Centre for Atmospheric Science. His research interests concern large-scale atmospheric dynamics and its impact on human and environmental systems. In 2012, he founded the energy-meteorology research group. He is involved in a wide range of academic and industry-partnered projects on weather and climate risk in the energy sector, covering timescales from days to decades ahead.

Marta Bruno Soares PhD, is a social scientist based at the Sustainability Research Institute at the University of Leeds. Her research focuses on climate services including the analysis of the science-policy interface, barriers and enablers to the use of climate information, and the value of climate information in decision-making processes. She is currently a PI on a Horizon 2020 (H2020) project looking at the development of climate services for agriculture in the Mediterranean Region and a PI on a Newton Fund Climate Science for Service Partnership in China project looking at the priorities for developing urban climate services in China.

Carlo Buontempo PhD, manages the Sectoral Information System of the Copernicus Climate Change Service at the European Centre for Medium-Range Weather Forecasts (ECMWF). He coordinates the activities of a large number of projects working on the interface between climate science and decision making in sectors ranging from energy networks to city planning. Buontempo completed a PhD in physics at the University of L'Aquila in 2004; then, he moved to Canada for his postdoc before joining the Met Office. Buontempo worked at the Hadley Centre for almost a decade where he led the climate adaptation team and more recently the climate service development team. In this role, he led numerous projects involving climate change adaptation and regional modelling in Europe, Africa, Asia and North America. In 2012 Buontempo became the scientific coordinator of EUPORIAS, a project funded by the European Commission to promote climate service development and delivery in Europe.

Steve Dorling PhD, is Professor of Meteorology in the School of Environmental Sciences at the University of East Anglia. After completing BSc and PhD degrees in 1992 he worked at Environment Canada as a visiting research fellow in the Long Range Transport of Air Pollution, before taking up a faculty position in Applied Meteorology at the University of East Anglia (UEA) in 1994. Complementing his academic position, Dorling co-founded the private sector company Weatherquest Ltd in 2001 where he holds the position of Innovations Director. Since 2015, Dorling has also been a Director of the World Energy and Meteorology Council. Dorling teaches meteorology at undergraduate level. Dorling is part of the Senior Management Team at UEA through his role as Associate

Dean in the Faculty of Science.   In 2013, Dorling co-authored the text *Operational Weather Forecasting*.

Laurent Dubus PhD, has been working with Électricité de France's (EDF) R&D since 2001 as an expert researcher in energy meteorology. He has skills and experience in climate system modelling, weather and climate forecasts and power systems management. His activities are dedicated to improving the effective integration of high-quality weather and climate information into energy sector policy formulation, planning, risk management and operational activities, to better manage power systems on all time scales from a few days to several decades. He is involved in different French and international activities and organisations at the nexus between energy and meteorology, including the World Energy and Meteorology Council (WEMC), the World Meteorological Organization (WMO), the Superior Council of Meteorology in France and the International Conferences Energy & Meteorology (ICEM) series. Laurent holds a PhD in physical oceanography.

John A. Dutton PhD, is the president of Prescient Weather, the chief executive officer of the World Climate Service, and a professor emeritus and dean emeritus at the Pennsylvania State University. He focuses on the analysis and mitigation of weather and climate risk in both private and public endeavours, including agriculture, energy, and commodity trading. He has a special interest in creating probabilistic climate variability predictions and scenarios as inputs for corporate decision systems and strategic planning. Dutton has experience in science and public policy, including the National Research Council, the National Weather Service, space and earth science, aviation and weather, and other environmental issues. He is a fellow of the American Meteorological Society and the American Association for the Advancement of Science.

Don Gunasekera is a research fellow with the Centre for Supply Chain and Logistics at Deakin University. His research interest lies in analysing issues along various supply chains including those across meteorology, infrastructure, and food and energy sectors. He has worked in a range of organisations including the Australian Bureau of Agricultural and Resource Economics, the Australian Bureau of Meteorology and Victoria University. During 2006–2009, he was the chief economist at the Australian Bureau of Agricultural and Resource Economics. He has written widely in domestic and international journals.

Sue Ellen Haupt PhD, is an NCAR (National Center for Atmospheric Research) senior scientist and Director of the Weather Systems and Assessment Program of the Research Applications Laboratory of NCAR. She is also Director of Education of WEMC and a Councilor of the American Meteorological Society (AMS). She previously headed a department at the Applied Research Laboratory of the Pennsylvania State University where she remains Adjunct Professor of Meteorology. She has also been on the faculty of the University of Colorado Boulder; the U.S. Air Force Academy (visiting); the University of Nevada, Reno; and Utah State University and previously worked for the New England Electric System and GCA Corporation.

Richard P. James PhD, is a senior scientist at Prescient Weather and the World Climate Service. He has a background in meteorological research, specialising in the high-resolution modelling of convective storms and in the application of modern meteorological datasets to problems of weather and climate risk management. James focuses on developing new techniques to empower weather-sensitive decisions, and he benefits from cross-disciplinary knowledge of scientific meteorology, statistics and finance. James received his BA degree in natural sciences from Cambridge University, and his MS and PhD degrees from the Pennsylvania State University.

Shylesh (Shy) Muralidharan is a global product manager with Schneider Electric DTN, focused on building real-time weather analytics solutions for energy applications. He believes that weather-based decision support systems will play a major role in making the future energy infrastructure smarter and climate-resilient. He has over 14 years of worldwide experience in product management and technology consulting in the energy and utilities sector specialised in strategy and solution design of smart grid technology projects. Muralidharan is a system design and management fellow from Massachusetts Institute of Technology (MIT) and has a bachelor's degree in mechanical engineering and an MBA from the University of Mumbai.

David Richardson PhD, is Head of Evaluation at ECMWF. He has over 30 years of experience in weather forecasting research and operations and has worked on all aspects of ensemble prediction methods for weather forecasts for weeks to seasons ahead. This includes the configuration of ensembles to represent the uncertainties in the initial conditions and modelling systems, development of products and tools for forecast users, and evaluation of forecast performance. He has published numerous scientific papers as well as book chapters on these topics. He is Chair of WMO Expert Team on Operational Weather Forecasting Process and Support, which oversees the co-ordination of operational NWP (numerical weather prediction) activities among WMO member states.

Jeremy D. Ross PhD, is Chief Scientist at Prescient Weather and Lead Forecaster of the World Climate Service. He has more than 15 years of experience researching and developing weather and climate models and innovative climate and weather risk products for energy, agriculture, retail, transportation, and the commodity markets. Ross has broad knowledge of the academic and private sectors, and that insight combined with extensive technical and analytical skills facilitates rapid development of innovative science for weather and climate risk management. Ross obtained BS, MS, and PhD degrees in meteorology from the Pennsylvania State University.

Emma Suckling PhD, is a postdoctoral research scientist within the Climate Division of the National Centre for Atmospheric Science in the Department of Meteorology, University of Reading. Her research interests are focused on climate variability and predictability, which includes interpreting and evaluating climate predictions, understanding the impacts of climate variability and change for energy (and other) applications, and extracting useful information from imperfect models. Suckling gained her PhD in the field of theoretical nuclear physics from the University of Surrey before making a transition into climate science, where she worked as postdoctoral research officer within the Centre for the Analysis of Time Series at the London School of Economics, before moving to her current role. She is also Chair of the Institute of Physics (IOP) Nonlinear and Complex Physics Group committee.

Alberto Troccoli PhD, is based at the University of East Anglia (UK) and is the Managing Director of WEMC. Troccoli has more than 20 years of experience in several aspects of meteorology and climate and their application to the energy sector, having worked at several other leading institutions such as NASA, ECMWF (UK), the University of Reading (UK) and Commonwealth Scientific and Industrial Research Organisation (CSIRO, Australia). Troccoli is the lead author of the UN's Global Framework for Climate Services' Energy Exemplar, the editor of three other books, including *Weather Matters for Energy*, and the convener of ICEMs. He holds a PhD from the University of Edinburgh (UK).

Robert Vautard PhD, is a senior scientist at the Centre National de la Recherche Scientifique (CNRS) and is working at the Laboratoire des Sciences du Climat et de l'Environnement (LSCE). He is a specialist in European climate and modelling of climate in relation to energy and air pollution. He was a review editor of the Intergovernmental Panel on Climate Change's (IPCC) Fifth Assessment Report (AR5), and co-authored 187 publications in peer-reviewed scientific literature. He is co-leading the energy branch of the Inter-Sectoral Impact Model Intercomparison Project (ISIMIP), an international project on impacts of climate change and theme of the WCRP (World Climate Research Programme) Grand Challenge of extreme events. He is leading the research and climate service activities of the Institut Pierre Simon Laplace (IPSL) excellence laboratory, and is a former director of LSCE.

LIST OF FIGURES

LIST OF TABLES

CHAPTER 1

Bridging the Energy and Meteorology Information Gap

Don Gunasekera

Abstract This chapter discusses the information gaps relating to the type, level of accuracy and frequency of delivery of specific weather and climate information, and what extra information is required by the energy sector in the coming years. It is argued that ongoing technical and scientific interaction between weather and climate service providers and the energy sector, supported by input from the information and communication technologies, can help bridge these gaps. This will help the users in the energy sector to both understand and respond appropriately to the available weather and climate information. Focusing on the linkages between weather-, climate- and energy-related information and data, the chapter draws attention to barriers to data sharing, benefits of overcoming the barriers and strategies to enhance data-sharing arrangements between the weather, climate and energy communities.

Keywords Service delivery • Partnership • Data sharing • Weather and climate information • Service providers

D. Gunasekera (✉)
Centre for Supply Chain & Logistics, Faculty of Science, Engineering & Built Environment, Deakin University, Burwood, VIC, Australia

© The Author(s) 2018
A. Troccoli (ed.), *Weather & Climate Services for the Energy Industry*,
https://doi.org/10.1007/978-3-319-68418 5_1

INTRODUCTION

Rise in global energy use has been modest over the past several years with growth rates of 1.1% and 1.0% in 2014 and 2015, respectively, and lower than its 10-year average of 1.9% (see International Energy Agency 2016; BP Global 2016). However, triggered by income and population growth, global energy use is expected to increase over the next several decades, particularly in emerging and developing economies. Weather and climate information is needed to efficiently plan, manage and operate energy services on very diverse space and time scales. Hence, weather and climate information is crucial, given the expected increase in global energy use in the coming decades and the interdependence of weather, climate and energy production and use.

World energy demand is estimated to increase by 48% from 2012 to 2040 (US Energy Information Administration 2016). Much of this increase in demand is expected among the developing non-OECD economies. Strong economic growth and expanding population in these economies will be the key drivers of rising energy use. Non-OECD energy demand is projected to rise by 71% from 2012 to 2040. In contrast, in the more mature energy-consuming and slower-growing OECD economies, total energy use is estimated to increase by only 18% from 2012 to 2040.

Given the estimated rise in world energy demand in the next several decades, the usefulness of weather and climate information for the energy sector will continue to be important and even increase. Hence, there is a growing demand for weather and climate information in the energy sector across many regions of the world. The practice of delivering weather and climate information requires sustained interactions with users along the energy supply chain. Both public and private sector meteorological service providers help to meet the weather and climate data requirements of the energy sector across the world.

A survey of clients of private meteorological service providers has revealed that the three main factors driving customer demand were the accuracy of their forecasts, assistance in operationalizing the forecasts and the availability of one-on-one consultation (Mandel and Noyes 2013).

The ability of weather and climate information providers to meet the growing requirements of the stakeholders along the energy supply chain has a range of gaps. For example, one of these gaps relate to the type, level of accuracy and frequency of delivery of specific weather and climate information, and what additional information is required by the energy industry in the near term and into the future. These gaps are not surprising in light of the complex interplay among the commercial, environmental, social and economic considerations involved in energy supply and the changing balance of energy technologies between fossil fuels, renewables and nuclear in different regions. Continuing technical and scientific interaction between weather and climate service providers and the energy industry (supplemented by input from other relevant sectors such as the information, communication and technologies [ICT] and the relevant regulators) can help bridge these gaps. This will enable the users in the energy sectors to both understand and respond appropriately to the available weather and climate information.

In this book, a range of gaps that hinder or slow down a more effective integration of weather and climate information in the energy sector business/decision making are covered. These gaps relate to, for example, (a) the increasing need for improvement in relevant weather forecast quality, (b) the growing demand for location specific and/or user specific and more targeted meteorological model outputs, (c) the continuing need for greater partnership between the energy and meteorological communities and (e) the emerging requirement to address data-sharing needs. In this chapter, these gaps are introduced and particular attention is focussed on data-sharing gaps in more detail.

FORECAST IMPROVEMENTS

Meteorological forecast improvements are an ongoing issue influenced by both supply- and demand-side factors. On the supply side, the accuracy of forecasts has been improving steadily over time. It is important to recognize that forecast improvements will depend on both improved model predictions and improved forecast formulation and delivery. Continuing advancements in technology and expertise in collecting observations, processing, analysing, making model predictions, formulating forecasts and disseminating them are some of the key supply-side factors enabling forecast improvements in relevant variables. Chapters 6, 7, 8, 9 and 10 of this book highlight various state-of-the-art methodological issues associated with short range, medium and extended range forecast improvements.

On the demand side, for example, from the perspective of the energy sector, the need for better and improved meteorological forecasts can arise due to a range of reasons, for example: to better manage energy production and distribution risks associated with weather and climate variability;

to efficiently operate energy markets and pricing; to undertake energy market regulatory compliance; to better plan and undertake the operation and maintenance of energy generation plants; to safeguard energy system assets from climate change impacts; and to develop risk management and adaptation planning. Several examples of these cases are covered in Chaps. 4, 10, 11 and 12 in this book.

According to the American Meteorological Society (AMS) (2015), opportunities for increasing forecast skill at all time ranges will need further research, close global cooperation and coordination, improved observations of the atmosphere, ocean and land surface, and the incorporation of these observations into numerical models. The avenues to improved model predictions include higher spatial resolution, more powerful supercomputers, wider use and improvement of model ensembles, the development of data mining and visualization methods that enable forecasters to make better use of model guidance and collaborative forecast development activities among operational forecasters and researchers.

Targeted Model Outputs

As highlighted in other chapters (see Chaps. 4 and 6, 7, 8, 9, 10, 11 and 12) in this book, location and/or user-specific higher resolution/downscaled/targeted meteorological model outputs can help address risks to the energy sector from, for example, extreme weather events, changes in water availability, unusual seasonal temperatures and rising sea levels. Hence, targeted spatial analysis and improved forecasts of mesoscale weather events are important for both short- and long-term energy system management.

Such targeted model outputs are also relevant in the context of rising share of renewables (e.g. solar, wind and hydro power) and the dependence of these renewable energy systems on weather and climate variability. Additional weather observations at strategic locations as determined by quantitative models are important in this context.

In regions where the share of renewables in the energy mix is expanding, relevant model outputs/forecasts targeted towards specific user groups such as transmission system operators (TSOs) would benefit considerably in integrating renewable electricity to the grid as also discussed in Chap. 5. Targeted weather decision support products also help system load forecasting, enhanced efficiency in pricing for hourly and bulk markets and energy market trading in general. Examples of the development of specific user targeted online interactive tools that can allow users to assess how energy production and demand will change in response to climatic factors in certain regions are provided in Chaps. 9, 10 and 12.

ENHANCED PARTNERSHIPS

Over the past several years, there has been an increasing incidence of research dialogues between energy and meteorological specialists, analysts and practitioners at domestic, regional and international levels. This promising development requires further enhancement in the form of regular bidirectional communications relating to issues such as improvements in current meteorological products and services used by the energy sector, new services and product requirements and generation, and skills and training needs of the users of relevant meteorological information. As reported in other chapters (see Chaps. 4, 5 and 13) of this book, specific user group oriented (e.g. wind energy or solar energy sector focussed) dialogues/workshops/meetings have enabled effective bidirectional interactions between, for example, meteorological experts and energy practitioners at technical, managerial and decision-/policy-making levels. The enhancement of the bidirectional interaction between energy and meteorology sectors will be underpinned by several factors. These include rising demand for meteorological services by the energy sector, continuing innovations in meteorological service development and provision, increasing availability of cost-effective digital technologies for service delivery, growing requests for development of codes, standards and guidance for meteorological information and emerging need to establish mutual trust among all stakeholders given that confidentiality issues can prevent energy firms from sharing specific meteorological requirements for operational practices in open (see Chaps. 4, 5 and 13).

DATA SHARING

The rest of this chapter expands more on data-related issues, firstly, because of their critical role and secondly, to provide specific indications about the factors associated with them. The linkages between weather, climate and energy are based on the fact that variation in weather and climatic conditions across short, medium, and long timescales can affect all energy sources and energy needs. Many public and private sector meteorological information providers fulfil the weather and climate data needs of the energy suppliers and users across many regions. In certain cases, some of these providers supply tailored weather and climate data to particular energy companies. Broadly speaking, an important challenge here is how to improve the interfacing of weather and climate data and information with key supply and demand-side activities of the energy industry. In this regard, sharing of relevant data is one area which is critical for enhancing the synergies within and between weather, climate and energy spheres. For example, improvement in data sharing within the energy industry could have benefits in terms of improving energy production/demand monitoring and forecasting. Also, enhanced sharing of weather and climate data could help minimize energy sector's vulnerability to extreme weather events, enable more cost-effective integration of renewable sources of energy and enhance energy supply and consumption strategies (American Meteorological Society 2012).

Kusiak (2015) argues that lack of data sharing in the renewable energy industry, for example, is hindering technical progress and limiting opportunities for improving the efficiency of energy markets. He points out that optimizing the supply of renewable energy requires data on device performance, energy output and weather predictions, seconds to days in advance. At present, large amount of these data is gathered by participants such as turbine manufacturers, operators and utility companies along the energy supply chain.

Pfenninger (2017) argues that energy research needs to catch up with the open-software and open-data movements. He points out several reasons why energy models and data are not openly available. They include business confidentiality, concerns over the security of critical infrastructure, a desire to avoid exposure and scrutiny, worries about data being misrepresented or taken out of context and a lack of time and resources.

Some competing energy companies also gather weather and climate data that they perceive as sensitive proprietary information. Hence, the amount of 'big' data within the meteorological sector and energy industry is rapidly growing. There is a growing demand in the energy sector to share its data openly so that better solutions for providing energy in a sustainable and cost-effective manner can be designed and implemented.

Often some private sector data is difficult for anyone outside to access without data-sharing agreements and non-disclosure arrangements. In this context, there are lessons to be learnt in relation to maintaining data confidentiality and security from other sectors such as commerce and health-care organizations (see Kusiak 2015). In general, non-disclosure agreements outlining the specifics of data sharing and results dissemination are used in data-intensive projects.

Barriers to Data Sharing

Meteorological and energy data-sharing arrangements are likely to be influenced by competition in service provision, commercialization of data, technical difficulties in data management, metadata problems, national security considerations and government data policies.

Better use of some data is often hindered due to data accessibility. Barriers to data sharing are generally based on the belief that it poses significant risks, from either side of the sectors. These risks may relate to identification of the businesses within datasets, misuse of the data resulting from misunderstanding of its quality or meaning, inappropriate exposure of commercially sensitive data and information and reputational damage due to release of information about the data custodians. Often lack of trust is also a key barrier to sharing data. Trust between data custodian and user or between custodians is essential in all circumstances. Many options, reflecting the nature of working relationships, can be used to build and retain trust as opportunities to access and share data expand (see Productivity Commission 2016).

Benefits of Data Sharing

Public sector meteorological service providers participate in regional, national and international data-sharing initiatives and obligations. They would not be able to perform many of their functions without the data exchange arrangements of the members of the World Meteorological Organization (WMO), which cover the public, private and research sectors of WMO member countries. The WMO provides an international framework through which its member countries coordinate the collection and exchange of information on the state of the global atmosphere, ocean and inland waters. The framework also supports the provision of essential meteorological and related services in all individual countries. The international exchange of essential data and products is free of charge under the provisions of Resolution 40 of the 12th Congress of the WMO. But Resolution 40 also places some restrictions on the commercial use of these data. Some shared meteorological data is only available under certain restrictions from the owner or provider. These restrictions may include a limited ability to use that data in situations other than those prescribed, or to charge for products/services derived from that data (see Australian Bureau of Meteorology 2016).

Increased accessibility of relevant meteorological data would likely enhance the forecast skill (and the underlying research and analysis) and facilitate integration of a wide range of energy resources. Sharing of data, while still safeguarding the commercial interests of individual data-providing private-sector companies, has considerable potential to benefit the entire energy sector. Data sharing and supply between meteorological service providers and energy industries require that arrangements are established, covering the types of data required, their frequency of delivery, the reliability of the service and targets for quality. This will require policies to promote sharing of data relating to meteorological services and energy demand among the research, forecast and operations communities; safeguards to protect the commercial interests of private-sector companies that share proprietary data and an enhanced data collection and quality-control capability for weather and climate observations (American Meteorological Society 2012).

In recent years, some public sector and some large search entities have supported a move away from releasing data under restrictive licenses to releasing data under more permissive 'Creative Commons' licenses that allow the data to be reused. Under the 'Creative Commons' licences, in general, others are allowed to use and distribute content as long as they credit the copyright holder. A 'Creative Commons' licence provides a simple standardized way for companies and institutions to share their work with others on flexible terms without infringing copyright. It allows users to reuse, remix and share the content legally. Offering one's work under a 'Creative Commons' licence does not mean giving up copyright. It means permitting users to make use of the material in various ways and under certain conditions (Productivity Commission 2016).

Enhancing the Data-Sharing Arrangements

Path to open-access data and hence data sharing could involve at least four basic elements: agreement to participation in a cooperative data-sharing regime, awareness of the problems and of the potential benefits of data sharing, formulation of data-sharing protocols and governance structures and development of data-and-knowledge sharing platforms (see Kusiak 2015).

Contreras and Reichman (2015) point out that to achieve widespread sharing of data, intellectual property, data privacy, national security and other legal and policy obstacles must be addressed. They observe four basic structural arrangements for scientific data pools (this may be applicable to weather, climate and energy sector-related data also) along a continuum ranging from the most to the least centralized. These include fully centralized, intermediate distributed, fully distributed and non-commons arrangements. According to them despite limited resources to link data repositories technically, there are advantages to fostering legal interoperability of data among distributed data repositories or custodians. To achieve this across different data users, rules for data access and usage must be compatible with each other, must comply with laws and regulations of relevant entities/jurisdictions and must address rights of ownership and control granted to data generators/custodians. The most straightforward path to legal interoperability is to contribute data to the public domain and waive all future rights to control it or to have data shared under standardized 'Creative Commons' licenses that have been widely used.

By using a risk-based approach to data access, custodians of data, whether public or private, could clarify and manage the nature of data risks. Risk could be assessed based on both the likelihood and probable consequence of data breaches. Where the potential implications of data breaches are non-trivial but likelihood is remote, custodians of data can still share or release, with mitigation strategies adopted as required. Also, access to the data needs to be carefully managed where the likelihood of breach and its consequence are considered high (see Productivity Commission 2016).

It is important to recognize that the private sector collects, stores and uses a vast amount of data (including weather, climate and energy sector data) and is almost certainly now the dominant controller of data in most economies. For example, retail energy utility sector generally has a small number of large firms, or even a single firm, serving a regional or national market. These utilities increasingly have the capacity to gather vast amount of detailed data on consumer energy use via sophisticated metering technologies such as 'smart meters'.

According to the Productivity Commission (2016), government intervention in support of data sharing or release may be warranted in certain circumstances such as insufficient business-to-business sharing and/or insufficient data/information released for the community benefit. Insufficient business-to-business data sharing may reflect monopoly holdings of data and misuse of market power. Having access to large quantities of data can give a company—particularly a large, vertically integrated business—a degree of market power. Such market power could be used to deter new entrants to a particular market. It is important to recognize that there are several mechanisms that allow business-to-business data sharing including bilateral commercial arrangements. It may still be the case that much wider access to some data could deliver greater public or community benefit. An important factor in considering these issues is the delivery of net benefits for the public while preserving commercial incentives to collect and add value to relevant data. Any mechanisms to increase access to privately held data would, however, need to be premised on a clear articulation of net benefits to the community and a demonstration that access to the relevant data is not able to be secured through other means including through existing private sector data marketplaces and platforms (see Productivity Commission 2016).

The European Union has mandated open access to electricity-market data, resulting in the creation of the ENTSO-E Transparency Platform to hold it. This highlights the fact that there are valid arguments for the creation of national energy-data agencies to coordinate the collection and archiving of a range of important data (see Pfenninger 2017).

Knowing the value of meteorological or energy data being shared is a key factor which can help assess the potential returns from adding value to such data by, for example, tailoring them for specific uses.

There are various approaches for valuing or pricing meteorological (see World Meteorological Organization 2015) or energy data. They range from free provision and marginal cost pricing to commercial pricing. The preferred approach will depend on user demand (willingness to pay for specific data) and the capability of the data supplier to act commercially. Most public sector meteorological data are currently available on an open access basis to enable full and free reuse. This can occur, for example, under an open access licence (such as Creative Commons Licence).

Where data is shared with other parties, the value placed on it will be determined by several factors: the availability of alternatives, the need for further processing for use, potential uses to which the data can be put and strategic leverage attached to the data (see Productivity Commission 2016).

REFERENCES

American Meteorological Society. (2012). The energy sector and earth observations, sciences and services. A policy statement of the American Meteorological Society. Adopted by the American Meteorological Society Council, 20 September. Retrieved from https://www.ametsoc.org/ams/index.cfm/about-ams/ams-statements/statements-of-the-ams-in-force/the-energy-sector-and-earth-observations-sciences-and-services/

American Meteorological Society. (2015). Weather analysis and forecasting. An information statement of the American Meteorological Society. Adopted by American Meteorological Society Council on 25 March. Retrieved from https://www.ametsoc.org/ams/index.cfm/about-ams/ams-statements/statements-of-the-ams-in-force/weather-analysis-and-forecasting/
Australian Bureau of Meteorology. (2016). Submission to the productivity commission inquiry into data availability and use, August. Retrieved from http://www.pc.gov.au/__data/assets/pdf_file/0019/206812/sub198-data-access.pdf
BP Global. (2016). Primary energy – 2015 in review. Retrieved from http://www.bp.com/en/global/corporate/energy-economics/statistical-review-of-world-energy/primary-energy.html
Contreras, J. L., & Reichman, J. H. (2015). Sharing by design: Data and decentralized commons – Overcoming legal and policy obstacles. *Science, 350*(6266), 1312–1314.
International Energy Agency. (2016). Key world energy trends. Paris: International Energy Agency. Retrieved from https://www.iea.org/publications/freepublications/publication/KeyWorldEnergyTrends.pdf
Kusiak, A. (2015). Share data on wind energy. *Nature, 529*, 19–21.
Mandel, R., & Noyes, E. (2013). Beyond the NWS: Inside the thriving private weather forecasting industry. *Weatherwise*, January/February.
Pfenninger, S. (2017). Energy scientists must show their workings. *Nature, 542*, 393.
Productivity Commission. (2016). Data availability and use, draft report, Canberra, October. Retrieved from http://www.pc.gov.au/inquiries/current/data-access/draft/data-access-draft.pdf
US Energy Information Administration. (2016). International energy outlook 2016. Washington, DC. Retrieved from http://www.eia.gov/forecasts/ieo/pdf/0484(2016).pdf
World Meteorological Organization (2015). Valuing weather and climate: Economic assessment of meteorological and hydrological services, WMO-No 1153, Geneva. Retrieved from https://sustainabledevelopment.un.org/content/documents/1972Valuing%20Weather%20and%20Climate%20Change.pdf

Open Access This chapter is distributed under the terms of the Creative Commons Attribution 4.0 International License (http://creativecommons.org/licenses/by/4.0/), which permits use, duplication, adaptation, distribution and reproduction in any medium or format, as long as you give appropriate credit to the original author(s) and the source, a link is provided to the Creative Commons license and any changes made are indicated.

The images or other third party material in this chapter are included in the work's Creative Commons license, unless indicated otherwise in the credit line; if such material is not included in the work's Creative Commons license and the respective action is not permitted by statutory regulation, users will need to obtain permission from the license holder to duplicate, adapt or reproduce the material.

CHAPTER 2

Achieving Valuable
Weather and Climate Services

Alberto Troccoli

Abstract Weather and climate services rely on the production and delivery of relevant, credible and, of course, valuable information. In this sense, the energy industry, with its long-standing and varied needs for these services and strong experience, provides a solid test bed for assessing these services. However, it is argued here that, whether for public or commercial use, weather and climate services are, in essence, no different to other more familiar services (e.g. financial). For weather and especially the more recent climate services to succeed, it is therefore important that lessons from these other common services—which also often deal with uncertain and complex information—are considered. It is also natural and important that the burgeoning climate services learn from the more mature and analogous weather services in order to leapfrog development. Initial public investment is critical to spur development of these services. Such investment should then be phased out in a managed way to avoid abruptly interrupting their growth phase and therefore jeopardising their sustainability, given the strong effort that is also being invested into developing these services. The criticalities of the weather and climate services—such as the accuracy and skill of the information—need to be borne in mind when modulating public investment.

A. Troccoli (✉)
World Energy & Meteorology Council, c/o University of East
Anglia, Norwich, UK

© The Author(s) 2018
A. Troccoli (ed.), *Weather & Climate Services for the Energy Industry*,
https://doi.org/10.1007/978-3-319-68418-5_2

Keywords Marketing approach • Innovation • Commercial value • Financial service • Stakeholder • user • customer • Transparency • Trustworthiness

WHAT'S A SERVICE—
NEVER MIND THE WEATHER AND CLIMATE?

Whether for public good or commercial use, the value of weather and climate information is ultimately measured in terms of its usefulness to society and, specifically, for the energy sector in our case. The way in which this information is conveyed can be multifaceted—it may be in the form of a temperature map for instance—but in order for it to be most effective it needs to be 'packaged' in terms of a *service*. Although the definition of *service* can vary depending on the objective, scope and maturity of the information or product, the overall aim of a *service* is to meet the requirements of the user of the service, by extracting the highest value from, in our case, weather and climate information for the specific application at hand, be it the forecasting of hydropower production or the impact of a snowstorm on the infrastructure used for electricity transmission.

Thus, in its most general sense, a *service* can be defined as:

A set of actions aimed at helping its beneficiaries make the best use of tailored information so as to improve their 'business'.

It is therefore apparent that a 'weather or climate' service is not unique, amongst other possible services. Although one can attempt a specific definition of weather or climate service,[1] it is useful to first try to understand how a weather/climate service differs from, say, a financial service, or from a medical service, or even a car service. These may look like disparate analogies but assessing these can help understand better what a weather/climate service is and what it is supposed to achieve. More specifically, we are arguing here that although each of these services delivers different outputs and outcomes, there is no fundamental distinction between them, in the sense of the above definition.

These services naturally differ from each other in terms of their specific features. If, for example, we characterise a service based on the following four features,[2]

- Maturity: how long have they been around?
- Tangibility: is it something we can easily relate to?
- Level of Risk: how reliable and/or accurate is the output/product?
- Trustworthiness (or Credibility): how much do we trust the service provider?

it is possible to categorise and inter-compare them as presented in Table 2.1. What this comparison tells us is that: (1) there is indeed nothing special about a weather/climate service from a conceptual point of view as they can easily be related/compared to other, more traditional and widespread, services; (2) weather and, especially, climate services carry a higher level of risk or 'caution' (e.g. of the likelihood or certainty of climate forecast) than the other comparator services. It should indeed be the focus of the experts involved in the development of weather and climate services to attempt to reduce these levels of 'cautioning' as much as possible. The next section provides some indication on how this *'caution' reduction* may be achieved.

PUBLIC VERSUS COMMERCIAL APPROACH—HOW DOES A SERVICE DIFFER IN THESE TWO CONTEXTS?

Before looking more closely at how the 'caution' level can be reduced, it is important to reflect a little more on the analogy amongst these services. One seemingly important feature, which was deliberately omitted in the comparison presented in Table 2.1, is the nature of the service. In other words, is the service commercial and/or a public good (including scientific) service? Although this feature appears to be critical, the reason it has

Table 2.1 Qualitative comparison between five different types of service based on the four representative features—Maturity, Tangibility, Level of Risk, Trustworthiness

	Maturity	Tangibility	Level of Risk	Trustworthiness
Financial Service	H	M	M	M
Medical Service	H	H	M	H
Car Service	H	H	L	M
Weather Service	H	M	M	M
Climate Service	L	L	H	M

The three qualitative levels—H, M and L—stand for High, Medium and Low, respectively. These are associated with colours to indicate the level of 'caution'. So, for instance, a high maturity (e.g. > 30 years) carries a low level of cautioning (green) whereas a high risk carries a high level of cautioning

been omitted in Table 2.1 is simple: the distinction between commercial (or private) good and public good, it is argued here, is essentially irrelevant if the aim is to deliver an effective service. Indeed, if a service is to be successfully adopted, it needs to be developed, promoted and implemented in a very similar way, irrespective of whether the service is commercial or public in nature. Indeed, while the way in which these three activities—development, promotion and implementation—is carried out may differ in the two contexts, they remain critical in both the commercial and public domains.

Another way to look at this is that lessons learnt in one domain can be transferred to the other. Thus, for instance, it is likely that a more commercially focussed 'traditional' sales approach can lead to an improved and more highly adopted public (weather/climate) service. Specifically, a sales approach would be based on the following four personal (i.e. human) characteristics or traits:

- Eloquence—To be able to influence the other person's decision about their need for the service
- Cultural awareness—To be able to relate to different people regardless of their backgrounds, gender, religion and so on.
- Flexibility—To be able to operate outside of day-to-day routine in order to better tackle new or unexpected situations
- Transparency and honesty—To critically believe in what one is promoting

Of the four traits, the last one is by far the most critical. If you cannot convince yourself that something is valuable and useful, having properly appraised its pros and cons, it is going to be very hard to convince others. In other words, you need to be genuinely convinced that the service is worth buying, and that you would actually buy it yourself—ideally, you have already bought it! And the best way to convince yourself is to be as transparent and honest as possible.[3]

While sales approaches differ, one should make the most use of the huge amount of knowledge accrued in the marketing arena (Aaker and McLoughlin 2010; Homburg et al. 2012), irrespective of the service or product for which these approaches were devised. And thinking that a public good service should be treated differently just because the user will not be directly charged—remember they would have already paid for it via taxation—can be a serious mistake, particularly if the service development process is led by non-commercially savvy people.

As remarked above, key to the uptake of a service is the careful development, promotion and implementation of a service. This process requires close interaction with the final users of the service. The aim of this interaction is to attentively capture the specifications of the service while also advising on the potentials a service can offer (e.g. see also Chaps. 3 and 4, for further discussion, especially in the context of the energy industry). Moreover, this process should be carried out in an iterative way—namely through regular consultations with the users. Overall, rigour and meticulousness need to be applied in the development of the service. Ultimately this approach will ensure the production of a useful, robust and long-lasting, and ideally replicable, service/product.

Adding Weather and Climate to the Service

Having drawn some analogies between meteorological (namely weather and climate) services and other more commonly known services, we here discuss how these analogies can be harnessed to produce better services for weather and climate. To a large extent, the discussion here prescinds from the source of funding required to develop meteorological services and from the quantification of the economic benefits of such services. Although these are very important discussions, they have extensively been authoritatively discussed in the literature (Freebairn and Zillman 2002a; Freebairn and Zillman 2002b; WMO 2015).

In spite of the close similarities between weather and climate services, there are three important distinctions to be drawn between weather services and climate services, leaving aside their intrinsic distinction according to which weather services essentially deals with forecasts of up to a few weeks in advance, and climate services deals with forecasts and projections from a few weeks to decades.[4] In terms of services, their main differences are:

1. Weather services are considerably more mature than climate services; the former have been around for 40+ years (Pettifer 2015), the latter have started to be developed in a consistent way only during the last decade (Hewitt et al. 2012);
2. Weather services are based on information (e.g. forecasts) that are both more accurate (shorter lead time) and verifiable (their lifespan is shorter, in line with its lead time) compared to climate information[5]
3. Data (including forecasts) policy for weather services can be different, and more restrictive, than that for climate services[6]

How these differences translate in practice is that weather services have accrued a strong basis, both in terms of products and (their related) market. From the low levels of the 1970s, European weather services now have an estimated value of €300 million per annum (excluding aviation) (Pettifer 2015). In addition, Lazo et al. (2011) estimated that US economic activity that is attributable to weather variability could be 3.4%, or $485 billion of the 2008 gross domestic product. It is therefore apparent that the weather services market—whether supplied by National Meteorological Services (NMS) or private sector providers—has a solid foundation.

This also means that users (or customers) of weather services are normally well aware of the potential and usefulness of weather products and therefore a smaller effort is required to persuade them compared to when climate products are promoted. For instance, there are established meetings for users, such as the European Centre for Medium-Range Weather Forecasts (ECMWF)'s 'Using ECMWF's forecasts' (UEF),[7] as well as a host of private weather service companies that are working to improve the customisation of weather products for their customers. While the cost of the basic weather information (typically the forecasts) charged by some NMS (as is the case for most European NMSs) has been an important barrier for start-up enterprises (Pettifer 2015), the market for weather services has been such that this barrier has been overcome by a large number of companies.[8]

Different is the case for climate services, for which the accuracy (or skill) of the product still plays a crucial limiting factor in their uptake.[9] A distinction needs to be made, however, between the climate forecasts, which have lead times from a few months to a year (also referred to as seasonal forecasts), and climate outlooks and projections, which consider time horizons from a few years to multi-decades ahead (referred to as decadal outlooks, over the following decade, and climate projections, beyond it).

In the case of seasonal forecasts, some regions of the world have some useful skill, namely they can be predicted more accurately than others (typically tropical areas), while others have little or no skill at all (normally regions at higher latitudes, like Europe, e.g. see Troccoli et al. 2008; Troccoli 2010), the skill being also dependent on the season and, particularly, the variable (air temperature usually has a higher skill than precipitation, or wind speed or solar radiation). It can therefore be difficult to convince a prospective user or customer to make use of such forecasts in areas or for variables that traditionally do not display an improvement in skill compared to using a long-term mean or climatology even if there is increasing evidence that their level of skill is improving (Alessandri et al. 2017). Thus, interest in and uptake of seasonal forecasts are increasing. Work on the applications of seasonal forecasts, initially spearheaded by the International Research Institute for Climate and Society (IRI),[10] is now becoming mainstream thanks to activities carried out at organisations such as the APEC Climate Center (APCC)[11] or programmes such as the Copernicus Climate Change Services (C3S).[12]

In the case of climate projections, it is essentially impossible to prove their level of accuracy. The best that can be done is to demonstrate the suitability of the climate models at representing features at the country or sub-country level (and not just global or regional averages). This can mainly be done on the basis of the climate model performance over the recent (observed) climate period.

From an application point of view, seasonal climate forecasts are relevant for operational matters such as resource management and infrastructure maintenance scheduling while climate projections are relevant for infrastructure planning purposes. Indeed, it is because of the increasing interest in these applications, with energy providing a prime example, that seasonal forecasts and climate projections are continually being developed and tailored to an increasing number of (prospective) users or customers (see also Chap. 12).

In spite of these advances, there remains the strong need for a close interaction with prospective users in order (1) to improve the service producer's understanding of the final use of the climate information so that it can be appropriately tailored and (2) for the user to better appreciate the strengths and limitations of the climate information. These are processes that require substantial time investment, both on the relationship building side and on the technical development side. Most important of all is the ultimate instillation of confidence in prospective customers in relation to climate services. This instillation does not mean over-selling the services, rather that one must extract the most information relevant for the specific application knowing the limitations of the product—remember the key marketing trait is 'Transparency and honesty' section 'Public versus Commercial Approach—How Does a Service Differ in These Two Contexts?'. Weather services have gone through this process much before climate services and so lessons could be learned from this experience, bearing in mind the fundamental difference between the two services, namely the level of accuracy and skill of the products. Lessons can also be learned by analogy from other services, as argued above particularly by adopting relevant commercial marketing techniques.

This type of approach is also supported by Brooks (2013) who argues that we now have the opportunity to provide innovative climate services (new datasets and products) to climate-sensitive sector clients. The key to achieving relevant, valuable and discontinuous climate services is to accelerate innovation in climate services, and decisively cross the Research-to-Operation valley of death. This, Brooks (2013) argues, can be achieved by adopting three Es—Engagement, Entrepreneurship and Evaluation.

It is sometimes argued that climate services should be mainly a public good exercise (e.g. Webber and Donner 2016). This view appears short-sighted since, as argued in this chapter, development of climate services can greatly benefit from trying to adopt a commercial approach. As these services mature, and the commercial value of climate services becomes more apparent, opportunities develop, and there will be an ever larger share of services that are offered at a cost. Thus the route to achieving a sustainable climate service is to embrace a standard, market-oriented approach, also expressed by the three Es of Brooks (2013). Of course, one should test whether it is more efficient, or cost-effective, to run a public good service (i.e. with public funding) or a commercial one. It can be easy to forget that public good services require substantial investment. This investment needs of course to be weighed against their benefit. More generally, while an initial public investment is beneficial, and even required, in order to set in motion a (commercial) activity, it is important to scale back public funding or subsidies (e.g. as done in the case of feed-in-tariff for solar photovoltaic installations in a number of countries), but in a mea-sured and reasoned way. One should also avoid falling into the trap of heavily funding a 'commercial' entity through public grants thinking that the services developed with these grants have value just because that com-mercial entity is involved in the development of the services. While such development may be useful to create capacity and eventually have a com-mercial entity which is self-reliant, it is entirely possible that reliance on public funding makes the 'commercial' entity risk-averse and therefore never really self-sustaining.

Summary

The development of weather and climate services should have as their ultimate objective the achievement of the best quality information to serve both the commercial and the public worlds. While weather services are well established, both in a commercial sense and in a public good way (e.g. through the weather forecasts regularly issued by media outlets), climate services can benefit from the lessons learnt by weather services, but also other analogous services (e.g. financial or medical services). Central to this analogy is the marketing approach and greater focus on the needs of specific customer groups, which at its heart requires an honest and transparent approach for it to be most effective in the long run. Having to deal with a product—climate information—which is highly uncertain and with relatively low skill, there is a strong need to create an environment of trust. Without this there is a risk to build a sand castle, which at the first high tide will be washed away.

Thus, the key to building valuable and self-sustaining (weather and climate) services is to grow confidence in the products by cultivating personal relationships with prospective users/customers. Achieving this objective requires substantial time and funding investment, particularly in the case of the newer climate services (compared to weather services). This investment should focus on providing strengthened education, training and knowledge transfer activities, as these are key components of enhancing confidence in a product/service.

Of course, investment to also improve and refine the underlying meteorological data and products is fundamental. Public funding that is being invested to develop these (climate) services, as in the case of C3S or the European Union H2020 programme, is therefore essential in this phase of development. Specifically, projects such as the European Climatic Energy Mixes (ECEM),[13] which is building a climate and energy demonstrator for the energy sector, are trying to make the best use of this funding by building a service for both commercial and public users, through a strong engagement with stakeholders.

The key question then is 'how much, and for how long, public investment should be used in order to achieve the best possible weather and climate services from both a public and commercial point of view?' While there is no easy answer, a good guide to addressing this question is to keep in mind the critical factors in the development of these services (e.g. accuracy and skill of products). Better understanding and analysing this key question could be an area of further investigation.

Acknowledgements The author would like to acknowledge two reviewers for valuable input to this chapter.

Appendix—Definitions of Climate Service

It is only recently that the climate community has got to grips with the concept of 'service'. As a consequence, this is still a relatively new concept, and that is perhaps why its definition is varied—here we present a few, verbatim, just to illustrate the point.

The Global Framework for Climate Services Definition

Climate services provide climate information in a way that assists decision making by individuals and organisations. Such services require appropriate engagement along with an effective access mechanism and must respond to user needs. Such services involve high-quality data from national and international databases on temperature, rainfall, wind, soil moisture and ocean conditions, as well as maps, risk and vulnerability analyses, assessments and long-term projections and scenarios. Depending on the user's needs, these data and information products may be combined with non-meteorological data, such as agricultural production, health trends, population distributions in high-risk areas, road and infrastructure maps for the delivery of goods, and other socio-economic variables.

http://www.gfcs-climate.org/what_are_climate_weather_services

The Climate Service Partnership Definition

Climate services involve the production, translation, transfer and use of climate knowledge and information in climate-informed decision making and climate-smart policy and planning. Climate services ensure that the best available climate science is effectively communicated with agriculture, water, health and other sectors, to develop and evaluate adaptation strategies. Easily accessible, timely and decision-relevant scientific information can help society to cope with current climate variability and limit the economic and social damage caused by climate-related disaster. Climate services also allow society to build resilience to future change and take advantage of opportunities provided by favourable conditions. Effective climate services require established technical capacities and active communication and exchange between information producers, translators and user communities.

http://www.climate-services.org/about-us/what-are-climate-services/

The Climate Europe Definition

A climate service is the provision of climate information to assist decision making. The service must respond to user needs, must be based on scientifically credible information and expertise and requires appropriate engagement between the users and providers. Policy makers can use climate services to access decision-relevant scientific information in order to make the best decisions for society as a whole. This can help society to cope with current climate variability and limit the economic and social damage caused by climate-related disasters. Climate services ensure that the best available climate science is effectively communicated with agriculture, water, health and other sectors, to develop and evaluate adaptation strategies.

https://www.climateurope.eu/definitions-climate-services/

NOTES

1. For climate services, several definitions exist, sometimes even circular; a few of them are presented in the Appendix.
2. Other helpful features could be identified, such as *usefulness of,* or *value of* or *share of the population affected by* the service, but the discussion here is not intended to be exhaustive, merely illustrative.
3. We say 'as far as possible' because there is no need for instance to disclose and discuss technical details, such as an approximation applied in a programming code used to produce a given map, as long as we genuinely think this is not affecting the overall message of the map.
4. The exact boundary between weather and climate services is sometimes blurred.
5. While seasonal forecasts can be verified, although not as frequently as weather forecasts, climate projections cannot, at least not in a strong statistical sense due to very long time horizon compared to the human lifetime and their limited sample.
6. While there are common data policies on common data sharing, data policies for weather forecasts differ substantially in say USA (free sharing) and Europe (data are charged). For further discussion, see Harrison and Troccoli (2010) or WMO (2015).

7. https://www.ecmwf.int/en/learning/workshops.
8. More needs to be done to completely remove or at least further reduce this barrier, however, as experience shows that an open-data policy tends to lead to a dramatic increase in the use of the data (Pettifer 2015, see also World Bank 2017).
9. Climate (and weather) services are here mainly treated in terms of predictions/projections. Historical (past) data could also be included but, while they might technically belong to climate services, they can be found in both weather and climate services.
10. http://iri.columbia.edu/.
11. http://www.apcc21.org.
12. http://climate.copernicus.eu/.
13. http://ecem.climate.copernicus.eu/.

References

Aaker, D., & McLoughlin, D. (2010). *Strategic market management: Global perspectives* (1st ed.). Chichester: John Wiley & Sons, 368 pp.

Alessandri, A., De Felice, M., Catalano, F., Lee, J.-Y., Wang, B., Lee, D. Y. et al. (2017). Grand European and Asian-Pacific multi-model seasonal forecasts: Maximization of skill and of potential economical value to end-users. *Climate Dynamics*. https://doi.org/10.1007/s00382-017-3766-y.

Brooks, M. (2013). Accelerating innovation in climate services. *Bulletin of the American Meteorological Society, 94*, 807–819.

Freebairn, J. W., & Zillman, J. W. (2002a). Economic benefits of meteorological services. *Meteorological Applications, 9*, 33–44.

Freebairn, J. W., & Zillman, J. W. (2002b). Funding meteorological services. *Meteorological Applications, 9*, 45–54.

Harrison, M., & Troccoli, A. (2010). Data headaches. In A. Troccoli (Ed.), Management of weather and climate risk in the energy industry. NATO Science Series (137–147). Dordrecht: Springer Academic Publishers.

Hewitt, C., Mason, S., & Walland, D. (2012). The global framework for climate services. *Nature Climate Change, 2*, 831–832.

Homburg, C., Kuester, S., & Krohmer, H. (2012). *Marketing management – a contemporary perspective* (2nd ed.). London: McGraw-Hill Education, 641 pp.

Lazo, J. K., Lawson, M., Larsen, P. H., & Waldman, D. M. (2011). U.S. economic sensitivity to weather variability. *Bulletin of the American Meteorological Society, 92*, 709–720. https://doi.org/10.1175/2011BAMS2928.1.

Pettifer, R. (2015). The development of the commercial weather services market in Europe: 1970–2012. *Meteorological Applications, 22*, 419–424.

Troccoli, A. (2010). Seasonal climate forecasting: A review. *Meteorological Applications, 17*, 251–268. https://doi.org/10.1002/met.184.

Troccoli, A., Harrison, M., Anderson, D. L. T., & Mason, S. J., eds. (2008). Seasonal climate: Forecasting and managing risk. NATO Science Series. Dordrecht: Springer Academic Publishers., 467 pp.

Webber, S., & Donner, S. D. (2016). Climate service warnings: Cautions about commercializing climate science for adaptation in the developing world. *WIREs Climate Change*. https://doi.org/10.1002/wcc.424.

WMO. (2015). *Valuing weather and climate: Economic assessment of meteorological and hydrological services*. World Meteorological Organization, 1153, 308 pp. Retrieved from http://www.wmo.int/pages/prog/amp/pwsp/documents/wmo_1153_en.pdf

World Bank. (2017). Open Government Data Toolkit. http://opendatatoolkit.worldbank.org

Open Access This chapter is distributed under the terms of the Creative Commons Attribution 4.0 International License (http://creativecommons.org/licenses/by/4.0/), which permits use, duplication, adaptation, distribution and reproduction in any medium or format, as long as you give appropriate credit to the original author(s) and the source, a link is provided to the Creative Commons license and any changes made are indicated.

The images or other third party material in this chapter are included in the work's Creative Commons license, unless indicated otherwise in the credit line; if such material is not included in the work's Creative Commons license and the respective action is not permitted by statutory regulation, users will need to obtain permission from the license holder to duplicate, adapt or reproduce the material.

CHAPTER 3

European Climate Services

Carlo Buontempo

Abstract The launch of the Global Framework for Climate Services (GFCS, Hewitt et al., *Nature Climate Change* 2(12): 831–832, 2012) just a few years ago helped to redirect the focus of the climate community towards the users and their information needs. A number of national and international initiatives such as the Climate Service Partnership, or the Climate Science for Services Partnership between China and the UK, were designed to build upon such an international framework. The role of the European Commission appears to be very prominent in the international climate services landscape as it supported a largenumber of research and innovation programmes in the field. The chapter discusses the role climate services could play for the energy sector starting from an analysis of the interactions that already exist and building upon a few specific examples that indicate some good practice in climate service development.

Keywords Climate services • Climate risk management • Copernicus climate change service • Climate information • Strategic investment • Energy trading

C. Buontempo (✉)
European Centre for Medium-Range Weather Forecasts (ECMWF),
Reading, UK

© The Author(s) 2018
A. Troccoli (ed.), *Weather & Climate Services for the Energy Industry*,
https://doi.org/10.1007/978-3-319-68418-5_3

27

Introduction

One of the characteristics that distinguishes science from services is the centrality of the users. Independent of the specific sector of application, the identification of users' needs is at the same time a complex and fundamental operation in climate service development (cf. Lemos and Morehouse 2005). The complexity of such a critical step is made even more complex by the loose understanding among users of what is and what is not within the scope of climate services (Lourenço et al. 2015).

What is clear is that before trying to identify their needs it is essential to identify who the users are. From a climate service perspective, a user is a person whose actions (e.g. decisions, policies) are likely to be influenced by the provision of a specific set of climate information. Such a definition makes it clear that a user is an individual operating in a specific environment rather than an organisation or a sector. This is quite different from having an industrial stakeholder within a research project. A stakeholder could represent the needs of an industrial sector, but it may not necessarily have an immediate decision to take. For a climate service provider a good connection with a company or a public administration can represent a necessary rather than a sufficient condition for the identification of a user.

The existence of a powerful public narrative around global warming and climate change (Lowe et al. 2006) can make the identification of the suitable decision-making person within the target organisation more complex rather than simpler. Especially the energy sector where the connection between climate change and energy production is extremely strong, it can be challenging to keep separate the discussion about the impact that climate variability and climate change could have on the business from the longer-term strategy of the company. The identification of a suitable decision-maker within the target organisation often represents one of the first challenges the development of climate service for the energy industry faces. This means that a direct contact between climate service providers and users is not common especially during the initial phase of development of a new service.

More generally, the simplistic model that sees a provider of climate information meeting a decision-maker and developing for them a well-defined service does not represent the norm. Whilst the user–provider connection can be more or less convoluted, in most cases it involves one or more intermediaries who transform a user-relevant but still general climate information into a product or a service which meets the requirements of a specific user.

What is critical to the success of the service is the ability to efficiently exchange information in both directions as user requirement can justify fundamental climate science development and users can see new opportunities once made aware of what the technology could currently achieve. A prerequisite for the development of a successful climate service is the establishment of a climate of mutual trust among all people involved (Brooks 2013).

Such a trust-building phase is necessary for a number of reasons. On the one hand, confidentiality issues can prevent industries from sharing specific requirements or operational practices in the open; on the other, a certain level of trust in the provider is required for users to consider the adoption of a specific service. Whilst it is at times possible to define sector-wide requirements such an assessment should be based on a generalisation of a series of user-specific requirements rather than a set of sector-wide requests which would otherwise be too loose and ill-defined to be action-able and acted upon. For example, whilst the energy sector as a whole may be interested in seasonal predictions of wind speed only, some specific professionals within the industry will use this information to inform specific decisions and they may require specific products (e.g. capacity factors at hub height or potential wind speed rather than 10 m wind speed as extracted from the models).

Energy Users' Requirements for Climate Services

Extrapolating from the experience acquired through EUPORIAS, a climate service project funded by the European Commission through its seventh framework programme (Hewitt et al. 2013), and the Copernicus Climate Change Service (C3S)[1], we can tentatively identify some general requirements that emerge from energy users. In general terms, climate information is currently used to manage environmental risks and to plan maintenance to critical infrastructure. Climate data is also one of the key inputs considered by energy traders. On the longer time horizon climate information also represents one of many inputs into strategic investment decisions. Figure 3.1 gives a schematic overview of these aspects which are presented in some detail in the following sections.

Climate Risk Assessment

From the dry soil conditions which can affect the heating dissipation (and thus the rating) of underground cables (Stern et al. 2003) to the wind, ice

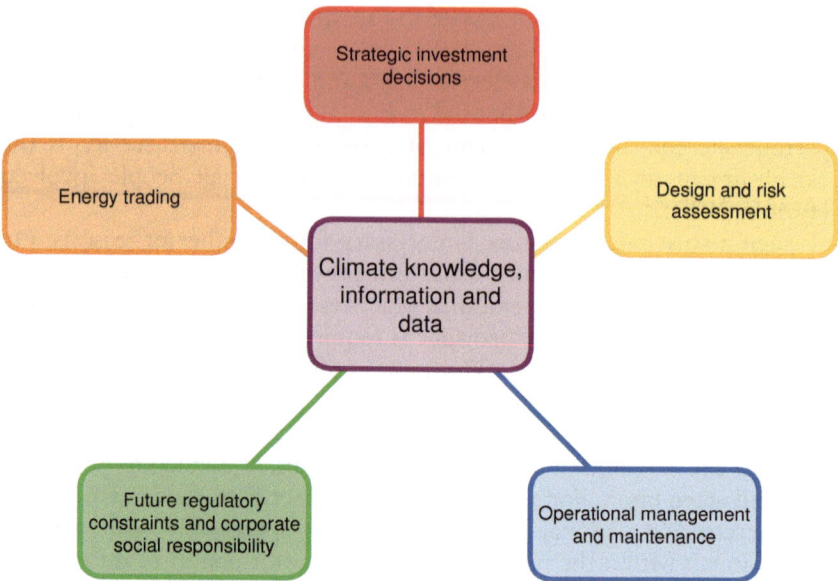

Fig. 3.1 A schematic representation of the ways in which climate information can be used within the energy sector

and snow storms which have crashed to the ground high-voltage transmission lines (Ward 2013; Campbell 2012), some of the aspects that have more directly affected the energy industry are directly related to the occurrence of isolated extreme climatic events or combination thereof.

In that sense, it is not surprising that one of the most common requests for climate information from Transmission Service Operators (TSOs) is related to climatological values (intensity and return periods) of extreme events in a changing climate. Civil and electrical engineers have built their practices on the assumption that the past weather events could provide a good guidance on the level of risk a specific infrastructure is likely to be exposed to. This is a solid approach in a world where the assumption of a stationary climate holds true, but it can become suboptimal if not even dangerous in a world characterised by a large, low-frequency climate variability and/or by a long-term trend in the parameters describing the statistical distribution of these variables (Wilby 2007). In general, long-lasting infrastructure requiring substantial investment is likely to be susceptible to a redefinition of building standards. However, in the energy sector, the expectations over such infrastructure are even higher given the impact and consequences that a failure in some of these systems may cause to society. For example, in the UK a nuclear power plant needs to be built so that it can withstand a 1 in 10,000-year return period flooding event (Starr 1981). It is thus clear that the analysis of environmental risks in a changing climate can become particularly complex for the energy sector (Rothstein and Parey 2011).

Strategic Planning

The second area of focus relates to the strategic planning of critical infrastructure. Traditionally, climate has not played a particularly big role in this kind of decisions, but the rapidity of the climatic changes and our growing capacity to predict some of them mean that this information is now playing a much more prominent role (Arnell and Delaney 2006; Larsen et al. 2008). For example, the viability of drilling and refining operations around the Caspian Sea (Zonn 2005) or the Persian Gulf will depend on a combination of freshwater availability, sea level rise and maximum daily temperature. Similarly, an investment decision in Floating Production, Storage and Offloading (FPSO) platform may be affected by the predicted change in wave conditions in the area as this can directly affect its design and ultimately the cost and possibly the return on investment of the infrastructure (Fonseca et al. 2010; Zou et al. 2014). Similarly, information on sea level rise or storminess could have a direct impact on the decision of decommissioning or not an offshore drilling platform (Burkett 2011).

In general, with the exception of renewable energy, climate information plays a relatively minor role in these strategic investment decisions. In fact, the economic, political and regulatory environment is likely to have more weight on the overall strategy that is adopted. A number of methodologies have been proposed to account for climate change and its associated uncertainty in investment decisions (Hallegatte et al. 2012; Lemos and Rood 2010) and an increase in the relative importance of climate information in strategic decisions in the years to come could be expected.

Corporate Governance, Planning and Communication

A recent analysis commissioned by the C3S through a contract led by the University of Reading has showed that one of the areas in which climate services are most used at the moment across all sectors is related to corporate governance and planning. Whilst the large uncertainty that comes with climate projections has often been identified as a barrier to the use of this information in strategic planning, having general indications, albeit uncertain, about what the future may hold appears to have a great relevance for corporate strategic planning. This appears to be particularly true for the energy sector possibly because of the direct link that exists between climate variability, climate change and energy. Given that legislation, public incentives and regulations have the power to affect the energy market significantly (e.g. Saidur et al. 2010), climate information could also be used by the industry to improve their understanding of how the regulatory framework may evolve in the years to come in response to societal pressure. For this kind of strategic decisions, information about mitigation strategies and carbon emissions are likely to be more relevant than the evaluation of climate change impacts. For example, having information about the likelihood to contain the global temperature change within 2 degrees from pre-industrial can have a significant bearing on the decarbonisation strategy the legislator is likely to pursue.

Operation and Management

The situation is rather different when looking at shorter timescales. Without necessarily entering the realm of weather predictions, it is clear that both historical climate and more recently climate predictions could be used to inform management decisions and operations (Troccoli 2010; Doblas-Reyes et al. 2013). This is of particular importance in a context where the fraction of renewable energy in the energy mix increases over time. Different from the traditional energy mix where climate represents simply an external factor, in the context of renewable energy the climate often represents the valuable asset itself. For example, planning the maintenance of an offshore wind farm can be an expensive operation which requires careful planning. Scheduling such a maintenance during a period characterised by relatively low wind conditions could both reduce the direct costs and reduce the loss in wind-energy production, as the turbine needs to be shut down during these operations. It is also becoming apparent that information about near-future conditions can provide useful insights into the return on investment. The 2015 wind drought in the USA has taken a lot of the industry by surprise and has impacted significantly the business plans of some of the operators. Although there is still some debate on the exact drivers of the drought, there is also evidence that it would have been possible to predict at least part of the observed wind anomalies.

Trading

Given its peculiarity in terms of climate information, energy trading deserves a category on its own. The highly interconnected nature of the European energy and the geographical disparities in terms of energy demand and production represent a good base for the energy trading in the old continent. As in any market, the operators tend to pay attention to as many different pieces of relevant information as possible. Given that both energy demand and (to a greater extent) energy production depend on weather conditions, it is natural that traders have always shown a great deal of interest in meteorological information. Very short-term (e.g. 0–48 hours) predictions, which are of greatest interest to traders, are now solidly in the hands of statistical post-processing algorithms (Foley et al. 2012). The trader forecasters are looking with growing interests at the predictions for the coming weeks and months as they feel that on those sorts of timescales their instinct and knowledge can still outperform the statistical tools and provide useful guidance for the traders.

The impact seasonal predictions have on the global gas market (Changnon et al. 1999) is an example of that and also shows how the market value of the predictions can often exceed the value climatologists would assign to it. Predicting something that is likely to affect the market is valuable per se even when the prediction itself turns out to be incorrect.

GOOD PRACTICE IN CLIMATE SERVICES DEVELOPMENT, FOR ENERGY AND BEYOND

One of the outstanding challenges of climate service development is related to the balance between user relevance/drive and public development. In Europe, where the European Commission has been investing heavily on climate services through both research programmes and innovation actions (Street et al. 2015), the challenge is becoming quite evident. On the one hand, developing a service without sufficient user engagement could lead to a product which is much closer to the providers' perception of the users' needs rather than something that is actually fit for purpose. On the other hand, publicly funding a service that only addresses the need of a specific user is also not in the interest of the taxpayer and it is almost certainly not politically viable. An example of the tension that may exist between these two opposing situations is provided by EUPORIAS. The project, structured around six climate service prototypes, focused on the climate prediction timescale. At least two of the prototypes were directly relevant to the energy sector: Hydrological Seasonal Forecast System (HSFS) to support spring flood regulation planning, which provides seasonal forecasts of the spring flood onset and volume (1–5 months ahead) in support of hydropower reservoir regulation planning in Sweden, and RESILIENCE, a user-friendly tool to produce information of the future wind power resources based on probabilistic climate predictions. In the case of HSFS, the prototype was entirely funded by Energiforsk, the Energy Research Institute of Sweden, which was also the target user for the service. It was also clear for which specific basin the prototype was going to provide information. In the case of RESILIENCE, the project team went through a series of stages to identify the target users.

Figure 3.2 provides a snapshot of the award-winning[2] graphical user interface they developed called project Ukko.[3] The development of such visualisation, which represented the graphical user interface to the data generated within the RESILIENCE prototype of EUPORIAS, also provides a good example of the kind of tension that may arise during the development of a climate services. The tension was in this case between EUPORIAS management team who were keen to develop a very targeted product addressing the need of a specific user (e.g. the manager of a wind farm) and the project team who were keen to develop a generic platform able to serve with relatively little modification a variety of users. On the one hand, there was the intention to understand how much the tailoring could add to the usefulness of a service. On the other hand, there was the cost-effectiveness requirement to invest in a system that could be reused for other applications. There is no reason to believe that this kind of tension, which ultimately represents a design challenge, might be a general issue of climate service development.

Opportunities for Climate Services, for Energy and Beyond

The Copernicus programme, previously known as Global Monitoring for Environment and Security (GMES), is a European system for monitoring the Earth System. It consists of a number of platforms which collect, process and distribute data from multiple sources such as satellite and in situ sensors. Both elaborated and raw data provide users with reliable and up-to-date information through a set of services, which address six thematic areas: land, marine, atmosphere, climate change, emergency management

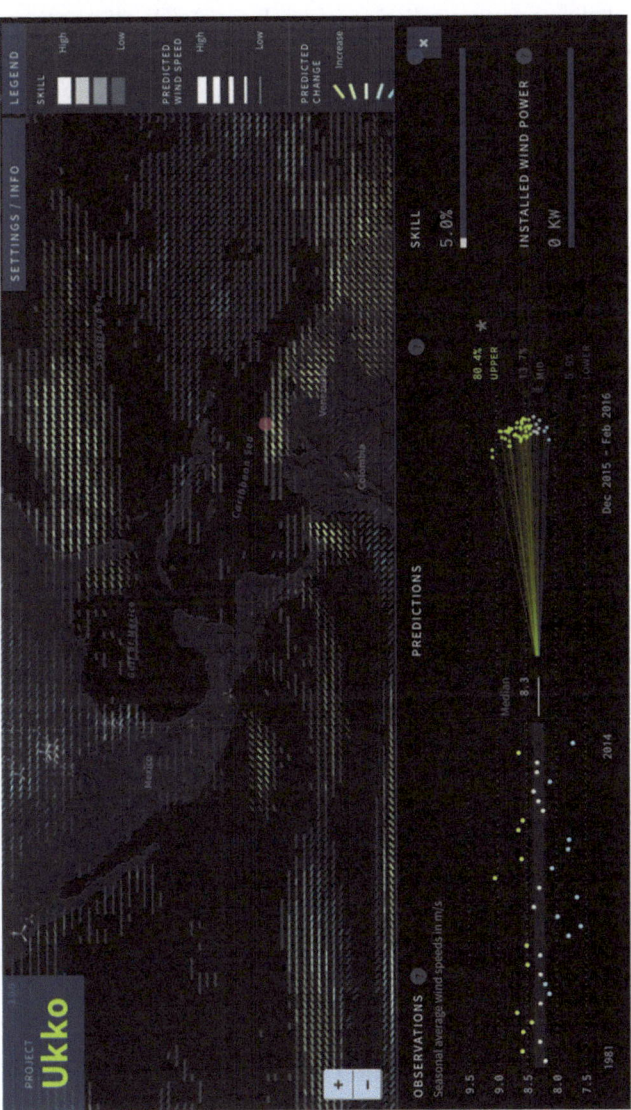

Fig. 3.2 Predictions of wind speed from ECMWF System 4 for the three months from December 2015 to February 2016 generated on November 2015. The colour of the glyphs and their directions encode the most probable category, that is, the tendencies of the ensemble mean of the seasonal forecast with respect to the model climatology at that location. The thickness of the glyphs indicates the mean wind speed predicted for the coming season. The opacity of the colour provides a measure of the skill of the prediction at that location measured by the Ranked Probability Skill Score. The regions with no glyphs are the regions where climate predictions for the selected months provide no additional information to the one available from climatology. When selecting a specific location, the user can see (bottom panel) the historical time-series for wind speed (bottom left) and the future predictions in the form of a probability cone (bottom right)

and security. The C3S, which is being implemented by the European Centre for Medium-Range Weather Forecast, is providing information for monitoring the global climate and predicting its evolution and will, therefore, help to support adaptation and mitigation efforts. The service will be built upon networks of in situ and satellite-based observations, re-analysis, seasonal predictions and climate projections. C3S will provide free and unrestricted access to several climate indicators and climate indices for both the historical and the future period.

The central piece of the C3S structure is the Climate Data Store, https://www.ecmwf.int/en/newsletter/151/meteorology/climate-service-develops-user-friendly-data-store. Despite its name, this is much more than a store of data as it represents a standardised access point to datasets (both those that exist already and the newly developed one) as well as a place where new application-relevant data will be made available. The data will include past observations and reconstructions, regional and global re-analysis, climate predictions and regional and global climate projections. The data will come with a standard set of tools which will allow users to define and then apply a range of post-processing procedures prior to the download of the data or graphics material they might need.

Alongside the development of the CDS, C3S is also developing a number of other important functions such as an Evaluation and Quality Control (EQC)—which will first define and then implement quality control procedures for all the dataset that will be made available by the C3S—and an Outreach and Dissemination (OD) function—which will be responsible for both training the users and the intermediaries and maintaining a support for the products and datasets.

One of the key aims of the C3S programme is to instigate the development of a market of climate services which could be built upon the free and unrestricted data policy of Copernicus. Although data accessibility is key to this vision, data on its own may not suffice for the uptake of climate information and data by users. To address this, the C3S is also developing a Sectoral Information System. This sub-programme is funding the development and the delivery of proof-of-concept demonstration services addressing the needs of specific sectors and users. The aim is to promote the development of conditions that will enable the use of the climate information provided rather than supporting a fleet of services for the end-users. Of the seven projects funded to date (Autumn 2016), two are designed to develop tools and datasets for the energy sector.

The first C3S Energy contract (European Climatic Energy Mixes, ECEM[4]), coordinated by the University of East Anglia, is looking at how different energy mixes will be able to meet demand on timescales ranging from the next season to the next decades in Europe. The main target of this C3S project is the development of an online interactive tool that will allow users to assess how energy production and demand will change in response to climatic factors in a specified region of Europe on different time horizons. This project, which at the time of writing is half-way through its completion, has already managed to identify a set of important technical developments that need to take place in order for the energy sector to be properly served. In addition, it has already provided their target users with something tangible they can play with: web-based demonstrator with up-to-date information about energy production and demand. Having a concrete tool to interact with is a fundamental step in the assessment of users' requirements.

The second C3S contract for the energy sector (CLIM4ENERGY[5]) is coordinated by the Commissariat á l'Énergie Atomique et aux énergies alternatives (CEA) and is expected to deliver nine energy-relevant pan-European indicators of climate trends and variability with cross-sectoral consistency, something that we believe will help users assess how exposed to climate extremes their infrastructure is likely to become in the coming decades.

Whilst none of these initiatives in isolation will be able to equip the energy sector with all the tools it needs for the challenges it is likely to face, these prototype service and demonstrators will provide useful examples for others to build upon. The recent adoption of Energy as a priority sector for the GFCS of the World Meteorological Organisation (WMO) means there is a general framework in which these experiences can be accounted for (WMO 2017).

Acknowledgement Project Ukko is a Future Everything and BSC project for EUPORIAS. Data visualisation by Moritz Stefaner. EUPORIAS is a project funded by the EU 7th Framework Programme (GA 308291) and led by the Met Office.

NOTES

1. http://climate.copernicus.eu/.
2. Project UKKO received the silver prize for the Kantar Information Is Beautiful Award http://www.informationisbeautifulawards.com/news/188-2016-the-winners.
3. http://www.project-ukko.net.
4. http://ecem.climate.copernicus.eu/.
5. http://clim4energy.climate.copernicus.eu/.

REFERENCES

Arnell, N. W., & Delaney, E. K. (2006). Adapting to climate change: Public water supply in England and Wales. *Climatic Change, 78,* 227–255.

Brooks, M. S. (2013). Accelerating innovation in climate services: The 3 E's for climate service providers. *Bulletin of the American Meteorological Society, 94*(6), 807–819.

Burkett, V. (2011). Global climate change implications for coastal and offshore oil and gas development. *Energy Policy, 39*(12), 7719–7725.

Campbell, R. J. (2012). *Weather-related power outages and electric system resiliency.* Washington, DC: Congressional Research Service, Library of Congress.

Changnon, D., et al. (1999). Interactions with a weather-sensitive decision maker: A case study incorporating ENSO information into a strategy for purchasing natural gas. *Bulletin of the American Meteorological Society, 80*(6), 1117–1125.

Doblas-Reyes, F. J., et al. (2013). Seasonal climate predictability and forecasting: Status and prospects. *Wiley Interdisciplinary Reviews: Climate Change, 4*(4), 245–268.

Foley, A. M., et al. (2012). Current methods and advances in forecasting of wind power generation. *Renewable Energy, 37*(1), 1–8.

Fonseca, N., et al. (2010). Numerical and experimental analysis of extreme wave induced vertical bending moments on a FPSO. *Applied Ocean Research, 32*(4), 374–390.

Hallegatte, S., et al. (2012). *Investment decision making under deep uncertainty – application to climate change.* World Bank Policy Research Working Paper 6193. Washington, DC: World Bank.

Hewitt, C., Mason, S., & Walland, D. (2012). The global framework for climate services. *Nature Climate Change, 2*(12), 831–832.

Hewitt, C., Buontempo, C., & Newton, P. (2013). Using climate Predictions to better serve society's needs. *Eos, Transactions American Geophysical Union, 94*(11), 105–107.

Larsen, P. H., et al. (2008). Estimating future costs for Alaska public infrastructure at risk from climate change. *Global Environmental Change, 18*(3), 442–457.

Lemos, M. C., & Morehouse, B. J. (2005). The co-production of science and policy in integrated climate assessments. *Global Environmental Change, 15*(1), 57–68. ISSN: 0959-3780. https://doi.org/10.1016/j.gloenvcha.2004.09.004. Retrieved from http://www.sciencedirect.com/science/article/pii/S0959378004000652

Lemos, M. C., & Rood, R. B. (2010). Climate projections and their impact on policy and practice. *Wiley Interdisciplinary Reviews: Climate Change, 1*(5), 670–682.

Lourenço, T. C., et al. (2015). The rise of demand-driven climate services. *Nature Climate Change*, *6*, 13–14.

Lowe, T., et al. (2006). Does tomorrow ever come? Disaster narrative and public perceptions of climate change. *Public Understanding of Science*, *15*(4), 435–457.

Rothstein, B., & Parey, S. (2011). Impacts of and adaptation to climate change in the electricity sector in Germany and France. *Climate Change Adaptation in Developed Nations*, 231–241 (Springer Netherlands).

Saidur, R., et al. (2010). A review on global wind energy policy. *Renewable and Sustainable Energy Reviews*, *14*(7), 1744–1762.

Starr, C. (1981). Risk criteria for nuclear power plants: A pragmatic proposal[1]. *Risk Analysis*, *1*, 113–120. https://doi.org/10.1111/j.1539-6924.1981.tb01406.x.

Stern, E., Newlove, L., & Svedin, L. (2003, January 1). *Auckland unplugged: Coping with critical infrastructure failure*. Lanham, MD: Lexington Books. ISBN: 9780739107744.

Street, R., et al. (2015). A European research and innovation roadmap for climate services. *European Commission*, 702151. ISBN: 978-92-79-44341-1. https://doi.org/10.2777/702151.

Troccoli, A., Boulahya, M. S., Dutton, J. A., Furlow, J., Gurney, R. J., & Harrison, M. (2010). Weather and climate risk management in the energy sector. *Bulletin of the American Meteorological Society*, *6*, 785–788. https://doi.org/10.1175/2010Bams2849.1.

Ward, D. M. (2013). *Climatic Change*, *121*, 103. https://doi.org/10.1007/s10584-013-0916-z.

Wilby, R. L. (2007). A review of climate change impacts on the built environment. *Built Environment*, *33*(1), 31–45.

WMO. (2017). *Energy exemplar to the user interface platform of the global framework for climate services*. World Meteorological Organisation, 120 pp. Retrieved from https://library.wmo.int/opac/doc_num.php?explnum_id=3581

Zonn, I. S. (2005). Environmental issues of the Caspian. *The Caspian sea environment* (pp. 223–242). Berlin; Heidelberg: Springer.

Zou, T., Jiang, X., & Kaminski, M. L. (2014). Possible solutions for climate change impact on fatigue assessment of floating structures. *The Twenty-fourth International Ocean and Polar Engineering Conference*. International Society of Offshore and Polar Engineers.

Open Access This chapter is distributed under the terms of the Creative Commons Attribution 4.0 International License (http://creativecommons.org/licenses/by/4.0/), which permits use, duplication, adaptation, distribution and reproduction in any medium or format, as long as you give appropriate credit to the original author(s) and the source, a link is provided to the Creative Commons license and any changes made are indicated.

The images or other third party material in this chapter are included in the work's Creative Commons license, unless indicated otherwise in the credit line; if such material is not included in the work's Creative Commons license and the respective action is not permitted by statutory regulation, users will need to obtain permission from the license holder to duplicate, adapt or reproduce the material.

CHAPTER 4

What Does the Energy Industry Require from Meteorology?

Laurent Dubus, Shylesh Muralidharan,
and Alberto Troccoli

Abstract The energy sector significantly depends on weather and climate variability, which impacts both demand and supply, at all timescales. Over the next decades, climate change mitigation and adaptation will lead to an overhaul in energy systems, to reduce greenhouse gases emissions. Low carbon energy generation is key to facing this challenge, but its renewable part—mainly from wind, solar and hydro power—will even increase the exposure of the sector to weather and climate factors. Energy companies can assess their preparation to tackle the impact of weather volatility on their operations by running a weather-readiness assessment. This chapter provides an overview of the energy sector today, together with future scenarios and challenges. The weather-readiness concept is then presented in detail and demonstrates that stronger collaboration between the energy

L. Dubus (✉)
EDF – R&D, Applied Meteorology group, Chatou, France

S. Muralidharan
Schneider Electric, Burnsville, MN, USA

A. Troccoli
World Energy & Meteorology Council, c/o University of East
Anglia, Norwich, UK

© The Author(s) 2018
A. Troccoli (ed.), *Weather & Climate Services for the Energy Industry*,
https://doi.org/10.1007/978-3-319-68418-5_4

industry and the meteorological community is key to reducing the risks posed by climate variability and change, and allow a more effective integration of high-quality weather and climate information into energy sector activities, to better manage power systems on all timescales from a few days to several decades.

Keywords Energy systems • Energy scenarios • Supply • Demand • Weather readiness

INTRODUCTION

The energy sector is weather and climate dependant. Both day-to-day weather and longer-term climate variability have impacts on supply, demand, transport and distribution, and energy markets. Despite the energy sector being one of the most advanced users of weather and climate information, its rapid evolution constantly creates new needs, which require a new paradigm for a more effective exchange of information between meteorologists and energy sector users. Scientific progress on its own is indeed not sufficient to increase the value of weather forecasts. Indeed, improving decision-making processes, and hence the value of meteorology, also demands improved communication and mutual understanding between energy and meteorology people.

In the last decade, a burgeoning number of sessions in Energy & Meteorology at various conferences (e.g. American Meteorology Society, European Meteorology Society, European Wind Energy Conference) started the process from the meteorology side, aiming to engage the energy side. The International Conference on Energy & Meteorology (ICEM) series is going one step beyond to support a bidirectional stream of communication. The third ICEM in Boulder in 2015 provided a platform for a seminar on Energy for Meteorologists, with three main objectives:

1. To provide meteorologists with an overview of the energy sector/ business
2. To enhance awareness of the importance of weather and climate for the energy sector
3. To help foster a dialog between both communities and to identify major challenges which should be addressed in a co-design approach in the coming years

This chapter summarizes the content of the seminar and suggests ways to improve and develop collaboration between weather and climate scientists on the one side, and energy practitioners on the other.

Overview of the Energy Sector/Business

Energy systems are the engine of economic and social development. As stated in the SE4ALL[1] 2014 report (SE4ALL 2014): '*Energy is the golden thread that connects economic growth, increased social equity and an environment that allows the world to thrive. Energy enables and empowers. Touching on so many aspects of life, from job creation to economic development, from security concerns to the empowerment of women, energy lies at the heart of all countries' core interests.*' On the other hand, the energy sector is responsible for the largest share of anthropogenic greenhouse gas emissions. Reducing this footprint on global climate demands increasing the share of low carbon technologies, as well as increasing energy efficiency, while the global energy demand will continue to rise. Before going into further detailed implications of these basic elements, we first present here a quick overview of the energy sector, with a special focus on the power sector.

As things are moving very fast in the world energy landscape, the reader should note that the figures given here are a snapshot taken at the time of writing. There are a lot of resources when one wants to look in detail at the state of world energy systems. The most relevant ones are listed at the end of this chapter. The World Energy Council (WEC),[2] the International Energy Agency (IEA)[3] and REN21[4] annual publications are among the most relevant to keep updated on the status of the energy sector.

Peculiarities of Energy Systems

Energy systems exhibit some common features, like other public good service sectors such as water and transport infrastructure for instance. First, they are capital intensive, with huge investments needed when one considers grid development or construction and operation of large production units. They are also characterized by long life cycles, on the lower order of 20 years for wind or solar farms, for instance, to as much as, or even more than, 60–80 years for heavy infrastructures, like large power plants, hydropower dams and transport networks. Last, but not least, energy markets are fragmented geographically, most generally at national level, and are sometimes subject to security issues.

Energy systems consist of a great diversity of generation sources, with very different characteristics in terms of:

- Technical aspects: size, net generation capacity, efficiency, reliability, operating constraints...
- Economic aspects: fixed and variable (operating) costs can vary a lot from one production means to another
- Regulatory aspects: CO_2 or other gas emissions limits, security rules, health and environmental impact regulations.

Energy systems are not isolated, but fully integrated in human activities, and closely linked to other sectors, in particular water. Water is indeed necessary for energy generation, either as the cooling fluid for some thermal power generation units or as the engine for hydropower generation. In addition, pumping, treating and moving water requires electricity. There are then strong links and dependencies between the energy and the water sectors, and, consequently, also with the food sector. This interlinkage is referred to as the energy–water–food nexus. It reflects the fact that there are competing uses of a common resource (water) for different human activities. Competition between energy generation, water supply and crop irrigation is already an issue in water scarce areas. It will become even more problematic in the next decades, with increased tensions on water resources due to climate change impacts (WEC 2016a).

As this chapter focuses on the power sector, we must add here some specificities of electricity as a commodity:

- Real-time balance between generation and consumption: as storage capacity is limited and/or very expensive, electricity cannot be currently stored on a large scale. Real-time balance between consumption and production must then be ensured in real time;
- Electricity demand is very variable in time, with characteristics varying among countries, depending on the uses of electricity for heating, cooling or any other application;
- Prices are very volatile: the electricity sector is now liberalized in most countries, and prices can fluctuate strongly according to trading opportunities on the markets;
- Natural monopolies: despite liberalization, network topologies impose physical constraints even on interconnected networks, as only limited amounts of energy can flow from one country to its neighbours;

- Technical complexity: there are difficulties in controlling load flows, interactions between generation and transmission (network congestion, blackout risks) and very diverse plant characteristics;
- Economic model: economic dispatch of production is based on increasing variable costs, the cheapest generation unit being called first. This implies that the marginal production cost of electricity increases with volume, contrary to most other commodities, and that a kWh is more expensive during a peak in load. However, in some markets, wind and solar energy are preferred, in line with CO_2 emissions reduction targets, and must be the first supply source in the stack, requiring more flexibility from other sources to account for their variability in time.

The Current Global Energy Picture

The IEA annual Key World Energy Statistics provides a regular overview on past trends, the current picture, and projections to 25 years ahead for energy production and consumption, as well as CO_2 emissions and energy prices. The 2016 edition (IEA 2016a) confirms past trends: globally, total primary energy supply and final consumption have constantly increased since 1973. Table 4.1 presents the main figures.

In 2014,[5] the share of energy sources in the global final energy consumption was roughly subdivided as shown in Fig. 4.1 (REN21, 2016).

Table 4.1   Main trends in energy supply and consumption and electricity generation from 1973 to 2014

	1973	2014	Comments
Total primary energy supply	6101 Mtoe[a] (70,955 TWh)	13,699 Mtoe (159,319 TWh)	• Relative decrease in oil, increase in coal and natural gas
Total final energy consumption	4661 Mtoe (54,207 TWh)	9425 Mtoe (109,613 TWh)	• Largest increase in China/Asia • Decrease in oil, increase in electricity
Electricity generation	6131 TWh	23,816 TWh	• Increase in nuclear, natural gas, renewables, • Decrease in oil

Source: IEA (2016a)
[a]1 Mtoe = 1.163×10^4 GWh

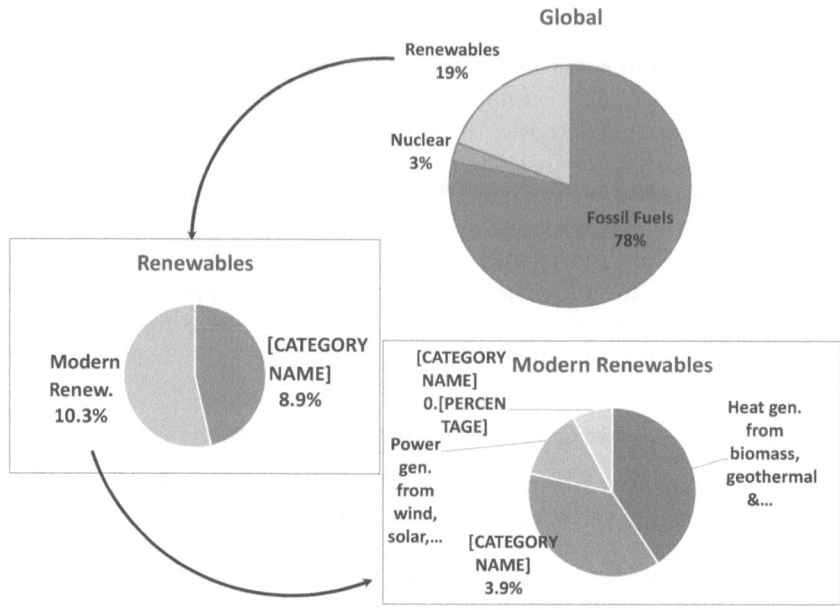

Fig. 4.1 Share of energy sources in the global final energy consumption (adapted from REN21 2016)

Obviously, fossil fuels still dominate at global scale. But this picture masks strong disparities between countries. For instance, whereas renewable sources represent only 8% of the final energy consumption in Belgium, this increases to 69.2% in Norway, mainly due to hydropower (WEC 2016b).

Focusing on power generation, wind and solar energy have been rapidly developing in the last decade, with an average annual growth of the installed capacity of 23% per year between 2004 and 2014 for wind and 51% per year for solar (WEC 2016c). This increase has been favoured by different factors, among which are incentives to develop these low-carbon-emitting technologies and a strong decrease in costs, in particular for solar panels. Overall, investments in renewable energy sources increased from US$72.8 billion in 2005 to US$285.9 billion in 2015, mainly dedicated to solar (US$161 billion) and wind (US$110 billion). As a result, the addition of new capacity in the power sector was higher in 2015 for renewables than for coal, gas, oil and nuclear combined (IEA 2016b). Overall, renewables, including hydropower, now account for about 30% of the total global installed power generation capacity and 23% of total global electricity production (WEC 2016c).

The effects of this energy transition towards cleaner technologies are becoming important also in terms of emissions. Indeed, CO_2 emissions from fossil fuels, which were increasing until 2013, seem to have stalled in 2014 and decreased in 2015 despite continued economic growth (Jackson et al. 2015, see their Fig. 1). The main reasons for this change in trend are a decrease in coal use in China, slower global growth in oil and the previously mentioned faster growth in renewables. The latter, together with a transfer from coal and oil to natural gas energy production, has implied a reduction in carbon and energy intensity, in particular in the USA and the UK.

Future Scenarios

Many organizations produce future scenarios for the energy sector, generally no further than 50 years ahead due to too many uncertainties. At institution level, a few scenarios are generally considered that cover a range of technical, economic and political options. Those scenarios are driven by the need for more energy supply, in response to increase in demand, and by the indispensable adaptation of the energy sector to climate change. Obviously, energy access for those who do not have secure, affordable and sustainable access to energy, the three pillars of the WEC Energy Trilemma (WEC 2016d), is an essential target. This currently concerns 1.2 billion people in the world. Despite differences between these scenarios, common features emerge. We here reproduce the main points highlighted by WEC (WEC 2016e) and IEA (IEA 2016b).

1. Energy demand will keep increasing, but per capita demand may peak around 2030 due to increased energy efficiency from new technologies and more stringent energy policies.
2. Fossil (coal, oil) fuels' share will decrease overall. Oil demand growth will be due mainly to freight, aviation and petrochemicals. Decarbonizing the global transport system is challenging, as the total number of vehicles is expected to double by 2040, and only few alternatives to fossil fuels exist. However, efficiency gains, use of biofuels and development of electric vehicles will make it possible to significantly reduce fuel demand for passenger vehicles. Among fossil fuels, natural gas will be more widely used.

3. Electricity demand is likely to double by 2060. This is driven first by the fact that the electricity sector will be easier to decarbonize than the others (transport in particular) and by an increase in urbanization and the associated development of technology-enabled lifestyles, which require more electricity.

4. Renewable sources, especially wind and solar energy, will be key in delivering low-carbon electricity. Today, they provide around 4% of power generation, but according to WEC's scenarios, their share could rise to between 20% and 39% by 2060 (WEC 2016e). Time and spatial variability of wind and solar generation poses risks to electricity security. Their integration will require a significant change in power system design and operation, allowing more flexibility to compensate for increases in variability from renewable generation. Several solutions therefore need to be developed in parallel to facilitate large wind and solar energy integration:

- demand response and management;
- increase in energy efficiency and conversion rates;
- stronger and smarter grids, including development of system services;
- availability of short-term backup power generation;
- energy storage, from water reservoirs, hydrogen, compressed air and batteries

The energy sector transition is now underway, and has been accelerated by the pledges made as part of the COP21 Paris Agreement in 2015. However, most energy scenarios show that the carbon budget for a 2°C target could be reached as soon as 2040 (IEA's main scenario, IEA 2016b), or between 2040 and 2060 (WEC 2016e). Most analysts therefore agree that the 2°C pathway will require stronger efforts than the currently pledged commitments, while a 1.5°C target has not yet been addressed from the energy point of view. Addressing the energy sector challenges will require global cooperation, sustainable economic growth and technology innovation, together with political decisions and actions to fix a (high) carbon price.

The Energy Trilemma

The WEC's definition of energy sustainability is based on three core goals—energy security, energy equity and environmental sustainability.

Balancing these three goals constitutes a 'trilemma' and is the basis for prosperity and competitiveness of individual countries. This trilemma neatly summarizes the energy sector's challenges. Since 2013, annual reports provide guidance to translate the three goals into tangible actions (WEC 2016d). More specifically, the three Energy Trilemma goals are defined as:

- **Energy security:** The effective management of primary energy supply from domestic and external sources, the reliability of energy infrastructure and the ability of energy providers to meet current and future demand;
- **Energy equity:** Accessibility and affordability of energy supply across the population;
- **Environmental sustainability:** Encompasses the achievement of supply and demand side energy efficiencies and the development of energy supply from renewable and other low-carbon sources.

The 2016 report (WEC 2016d) suggests five focus areas to achieve those goals, as shown in Fig. 4.2. Meteorology, that is weather observation, forecasts and longer-term climate projections, will play an increasing role in helping achieve these objectives. A more effective integration of weather and climate information into energy systems will also require

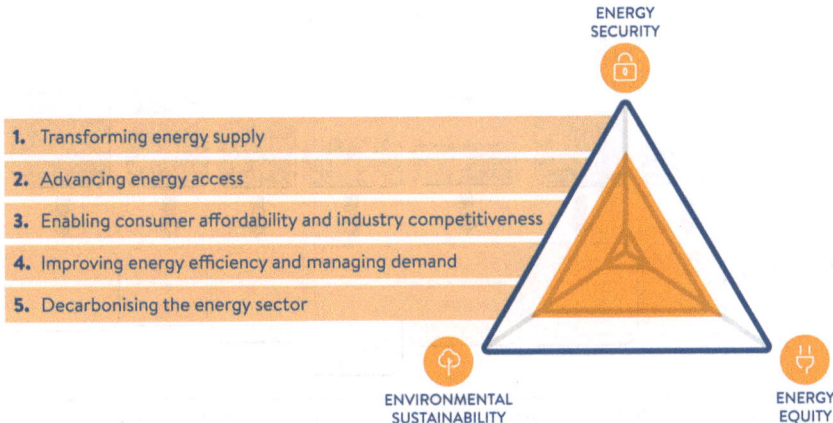

Fig. 4.2 The World Energy Council's Energy Trilemma, and the five focus areas for achieving energy goals (WEC 2016d). Used by permission of the World Energy Council

fostering the collaboration between the meteorology community and the energy sector, from the utilities level up to the policymakers level (WEC 2016d).

The Importance of Weather and Climate for the Energy Sector

Weather and Climate Impact the Energy Sector on All Timescales

Influences of short-term weather variability and longer-term climate impacts on the energy sector are well documented today. Among the available publications, most chapters, if not all, in Troccoli et al. (2014) give multiple examples and figures about these close links, which concern more or less all the fields of activity in the energy business, as exemplified in Fig. 4.3. In addition to the physical links between, for example, variable cloud cover and PV generation variability or between long droughts and reduced hydropower generation, weather and climate variability also have strong impacts on energy markets and energy system finance.

Variability of the weather-driven part of energy demand and supply causes energy prices to vary. The way energy prices vary depend also on

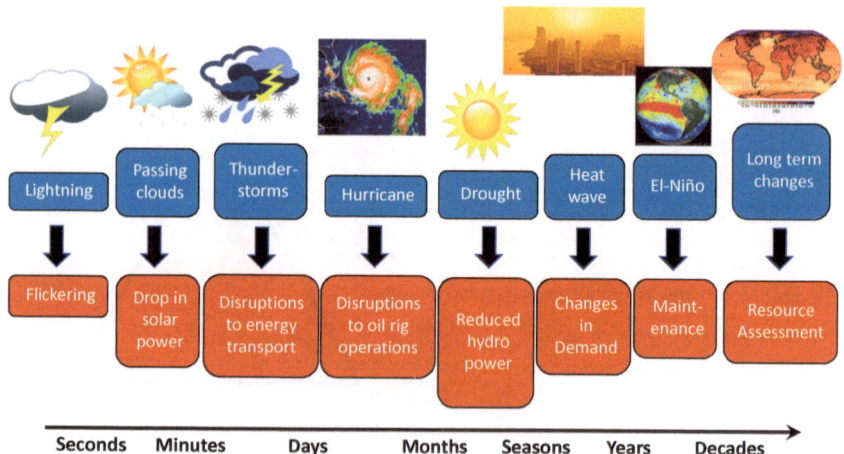

Fig. 4.3 Weather and climate impact the energy sector on all timescales (source: WEMC)

the underlying trading schemes. Long-term contracts are managed on forward markets, where week-ahead, month-ahead, quarter-ahead and even (up to three) year-ahead products are most often traded in the form of bilateral contracts.

Energy volumes can be purchased or sold on intra-day and day-ahead spot markets, similar to stock exchanges. In addition, in countries where energy load balancing is actively managed, Transmission System Operators (TSOs) and national regulators have set up balancing mechanisms to be able to mobilize extra generation or consumption reduction in case of higher than expected demand, and generation modulation or even curtailment in the case of lower than expected demand, the final goal being of course to ensure the supply–demand balance in real time. As electricity prices follow the offer/demand law, prices can increase significantly during peak demand. To fulfil low-emission energy development targets, renewable energy is prioritized in most countries to meet the demand. The increasing share of variable wind and solar generation then increases the volatility of net demand, defined as consumer demand minus renewable supply, and then that of prices. Any actor on the market who is able to forecast supply and demand better than competitors is then in a favourable position to buy and sell energy in a profitable way.

Energy system long-term financing is also more and more impacted by weather and climate, for two main reasons:

1. the increase in strongly weather-dependant wind and solar energy, and
2. the increasing vulnerability of energy assets to climate change.

Indeed, renewable energy projects' bankability (likelihood to ensure the financial success of the project), and hence financing through bank loans, depends on the projected resource and profit over the full investment and operation period, generally between 15 and 30 years. The resource estimation needs of course to be as accurate as possible, as any overestimation will result in less profit. Together with the technical choices about the production unit's characteristics, the resource estimation is a key factor in determining the Levelized Cost Of Energy (LCOE) of a project, this parameter often being the main or even the only determining selection criterion among competitive bids. As an example, EDF R&D and HYGEOS (Elias et al. 2015, 2016; Garnero et al. 2016) are developing new methodologies to estimate the solar radiation attenuation by aerosols in the surface layer (between 0 and 200 m AGL) in solar thermal plants, on the slant path from the heliostats (mirrors) to the concentrating tower. For instance, in Ouarzazate, Morocco, slant path attenuation can vary between 0% and 20% during desert dust events, significantly impacting the estimated long-term resource. Improved horizontal attenuation measurement and long-term estimation can significantly modify the LCOE calculation. This requires improvements in attenuation measurements during field campaigns in rough terrains, and new methodological developments to extrapolate short in situ measurements on the longer term, by coupling with satellite data and/or reanalysis.

More recently, long-term investments have started to take into account also the potential future impacts of climate change. As scientific consensus on future climate change has now been reached (IPCC 2013), this dimension has become a key factor in energy system development and new power plant projects. This is particularly true for long-term assets such as big dams or nuclear power plants, as well as power networks.

The International Hydropower Association[6] (IHA) status report 2016 (IHA 2016) notes that many countries are seeking a better understanding of climate change impacts and are beginning to build climate adaptation strategies and climate resilience into their plans. Many policymakers and industry leaders require guidelines and a robust framework for approaching climate risks. The World Bank is also very active in defining and developing guidelines for the hydropower sector (and more generally the whole energy sector), in order to help countries and businesses building resilience for both existing hydropower infrastructure and future projects. Its Energy Sector Management Assistance Program[7] (ESMAP) in particular provides analytical and advisory services to low- and middle-income countries to increase their know-how and institutional capacity to achieve environmentally sustainable energy solutions for poverty reduction and economic growth. In addition, many funding agencies and development banks now require that any long-term project includes a comprehensive assessment of climate change impacts (see for instance HRW 2014).

Many energy companies already consider climate change in their investment plans. For instance, EDF takes climate change into account when assessing the possible changes of efficiency of its thermal and nuclear power plant cooling systems (Anderhalt 2015).

Generally, the energy sector is, today, well aware that weather and climate variability and change impact their business, at every timescale. But not all companies and policymakers do take this quantitatively into account in planning and operations. The next paragraph shows why weather readiness is important, and how this concept can help the sector deal with weather and climate risks.

Weather Readiness Is Key for Weather-Resilient Business Performance for Electric Utilities

Weather-Readiness Assessment—Background and Introduction
Recent research (Francis et al. 2014) shows that in the past couple of decades, changing climate trends have led to increasing volatility in weather patterns across the world. This has significantly impacted the daily business operations of entities across the electricity supply value chain. Electric power generation entities are stressed during extremely anomalous hot or cold weather events as extreme weather forces them to 'make or buy' energy worth millions of dollars. Energy becomes a scarce commodity during extreme events and good knowledge of weather forecasts helps companies to take decisions to produce or procure energy at the right time and at an economical cost and help them be ready for severe weather in the most efficient manner. For companies involved in activities down the value chain such as electricity transmission and distribution, weather definitely has had an increasing impact on operational reliability (IDC 2013) in recent decades. Every time there is a severe weather event, the delivery of electricity is impacted and customers may be without power for several hours or days depending on the strength of the storm and the ability of the utility to restore power in their service territory.

Outages have a significant economic impact as well. Severe weather-related outages cost the US economy a total of approximately $80 billion annually (LaCommare and Eto 2006), half of which impacts the industrial and digital economy. Electric Power Research Institute (EPRI) research shows that the economic cost of power outages is largely related to the length of the outage, while noting even short duration outages of a few minutes could have large costs. It is estimated the average cost of a one-hour outage for manufacturing and digital economy firms is $7795/firm (Baggini 2008). Among industrialized countries, the USA has one of the highest annual average outage durations per customer, and also one of the highest annual average number of supply outages per customer; more than 45% of US utilities have a System Average Interruption Duration Index (SAIDI) metric of greater than 100 mins/year. The economic cost of these outages impacts not only end-users of utilities but also utilities themselves. More than 60% of utilities in the USA lose more than $100,000 on average per year in revenues due to outages and this number does not even include the unplanned costs of response and restoration.

Rationale for Investing in Weather-Readiness Assessment

So how do weather-readiness assessments help mitigate these problems? Weather-readiness assessment is a formal method which makes use of analytical tools to critically assess a utility's preparation to tackle the impact of weather volatility on its operations. This includes a thorough study of the current 'state-of-the-system' with respect to weather volatility and identifies specific actions to ensure operational resilience during severe weather. Using the weather-readiness assessment framework helps utilities identify the overall influence of weather, gaps in the current health of their systems with respect to weather resilience, and measure how weather variables impact their business goals of providing reliable power, maintaining service levels and reducing operational costs. There are several benefits to utilities adopting weather-readiness assessments. At an operational level, being weather-ready improves process efficiency and enhances service quality to the end-users and eventually improves customer satisfaction. At the larger industry level, electric utilities that undergo this assessment pioneer best practices in the sector and can utilize the outcomes of the assessment as a competitive advantage, by enhancing performance and cutting down costs associated with being weather-resilient. All utilities embarking on such an assessment send a positive message on how prepared the electricity sector is to handle the increasing trend of weather volatility. Finally, being weather-ready means that utilities are not 'reacting' to every weather event but instead have clarity of how weather impacts each of their functions and proactively figure out what they need to do to keep their operations resilient during storms and/or other severe weather events.

How Does the Industry Benefit from Being Better 'Weather-Ready'?

The severe weather volatility seen in different parts of the world in the recent past have forced utilities to make unplanned expenses in restoration and recovery efforts, especially during superstorms. These ad hoc expenses, especially when they are a significant investment during a multi-storm year, make their financials look less than optimal. Unplanned expenses can also divert funds from future investments leading to ageing infrastructure that has been denied the appropriate upgrades required for being weather-resilient, making it more vulnerable in the future, thus fuelling a vicious circle. So utilities have a good reason to design a weather-readiness assessment framework that will make them better prepared and able to respond effectively to the next severe weather event which will test their service reliability.

Defining Outcomes of Weather-Readiness Assessment
Weather-readiness assessments can be done in any part of the electric utilities value chain—whether it is electricity generation, transmission and distribution or retail services—but the assessment can be successful only if it is linked to business outcomes in the corresponding business. And for companies that operate in more than one part of this value chain, the assessments too can be modified to reflect a multitude of business outcomes (see Fig. 4.4). For example, at an enterprise level, it is a high-level goal to optimize the cost of emergency operations. At a 'generation' utility level, outcomes are linked to incorporating distributed generation (mostly renewables) effectively in the overall generation portfolio. On the transmission and distribution side, business outcomes are linked to bringing

Fig. 4.4 Business outcomes driving weather-readiness assessment. Electricity Value Chain Graphic adapted from 'Utility Analytics Market & Energy Analytics Market Global Advancements, Business Models, Worldwide Market Forecasts and Analysis (2013–2018)'

down the duration of outages during a storm, and finally, in retail it is linked to improved customer satisfaction by improving communication during storms and delivering superior service.

Once business outcomes are identified, they are converted into realizable operational outcomes. A risk management policy document that includes a section on ensuring adequate storm preparedness and mandating that all departments include weather readiness as part of their annual operational due diligence is one manifestation of an operational outcome. Other manifestations could support a heightened level of current situational awareness by using equipment and decision-support applications to get a real-time 'state-of-the-system' view of utilities' assets integrating climate and weather information. Sometimes it could also be linked to a future objective, so when there is a plan to install a new asset, a weather-readiness assessment could become an integral part of the due diligence of the location. That is the reason operational outcomes have to be backed by measurable quantitative metrics. These are metrics which will have to make economic sense for the utility, so an operational outcome focusing on improving service reliability will have to go through a cost-benefit analysis of reduction in the number/duration of weather-related outages juxtaposed with the cost to successfully achieve these metrics. It is also important that these metrics have the buy-in from various functions of the company as well as supporting comparison studies industry-wide. So in the above example, being reliant upon industry standardized service reliability metrics such as SAIDI and System Average Interruption Frequency Index (SAIFI) will add much more credibility to the assessment outcomes.

Weather readiness will have to be performed keeping expenses in check, so it is important that utilities do not embark on theoretical assessment studies that are a disproportionate sink of time and effort if they do not contribute effectively to making a good business case. And to make a business case, the true cost of the assessment must be taken into account, some of which could be indirect as well. An example of a direct cost could be the expense of integrating weather-based decision-support features into an existing utility distribution automation application, whereas an example of an indirect cost could be the opportunity cost of setting aside an exorbitant annual 'reserve fund' for multiple severe weather events and thereby denying much-needed system hardening investments within the utility's asset network.

Preparation for an Effective Weather-Readiness Assessment Framework
Once the weather-readiness assessment activity has been tested for its economic feasibility, the next step is to ensure that the data used for such assessments are robust. Assessing weather readiness as part of siting a new infrastructure within the service territory will require multi-year normalized climate data to understand weather patterns in a specific geographical location, while a situational awareness application focused on supporting severe weather events might need highly granular sub-second weather records such as lightning data that is updated in real time. Preparing the right kind of weather data also offers opportunities to coordinate and optimize requirements across multiple functions in the organization. So the power generation side of a bundled utility might be archiving weather data for forecasting load for several years and most of that archived weather data for multiple locations can also be used on the operational side to support machine learning algorithms to understand how specific weather variables such as temperatures and wind speeds play out in that location during extreme weather conditions.

Interesting Applications at the Intersection of Energy and Meteorology
Having a robust data layer allows utilities to design weather-based decision support across the utility. And decision-support applications can be delivered with varying levels of sophistication. At the simplest level, it could be a simple report which elaborates weather variables observed at any given location at a given time in the past week or month or for any given severe weather event. A post-mortem analysis of a heat wave or cold wave usually just requires a simple time-series of weather variables (usually of just temperatures) to understand the times when the utility demand was most stressed.

This might not be the case in an application which is a sophisticated decision-support application predicting asset damage from an incoming storm event. The same weather data in the earlier application is now regressed with data regarding location and performance of utility assets during similar weather conditions in the past. This 'historical patterns' data is then combined with non-weather variables such as vegetation and land use data to identify how a utility service territory will hold up during forecasted stormy weather based on historical behaviour. This allows utilities to predict and simulate severe weather's impact on their facilities and predict service interruptions to their end-users. The same data can also drive automated applications which can optimize system decisions such as reconfiguring power through alternative transmission or distribution network paths based on availability of network bandwidth which is not impacted by an incoming storm. This ensures that the degree to which service level performance is compromised is kept to a minimum and customer satisfaction levels can be maintained by constant communication and prompt restoration.

At a retail service level, products such as smart thermostats and services and smart utility bills normalized using weather data utilize hyperlocal weather data at a location to deliver unique decisions for energy management of residential, commercial and industrial premises. For a residential customer, it aids lifestyle decisions for the overall comfort while for a commercial customer it supports energy-efficiency initiatives. Of course, as the electrical grid gets smarter, the possibilities of how weather data can be applied using electricity supply and demand data at multiple nodes in the power grid can be endless. Figure 4.5 shows examples of current and future electric utility applications (across enterprise, grid and consumer applications) which can be enhanced by weather-based decision support.

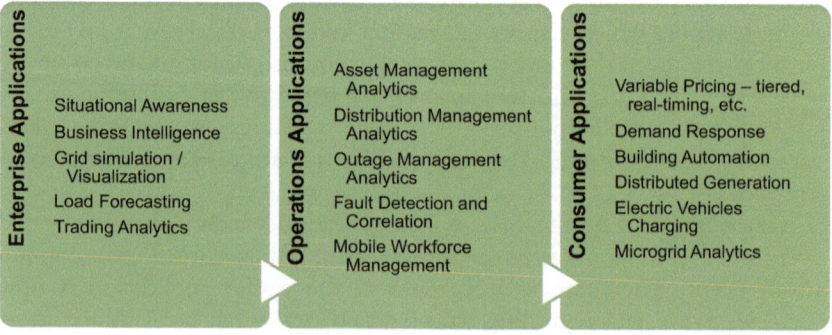

Fig. 4.5 Electric power sector applications enhanced by weather-based decision support. *Graphic Adapted from 'Utility Analytics Market & Energy Analytics Market (Solar Analytics, Oil & Gas Analytics, Water analytics, Waste analytics): Global Advancements, Business Models, Worldwide Market Forecasts and Analysis (2013–2018)'*

Even though weather-readiness assessments can follow a generic framework, they have to be customized in a specific way for any utility to take advantage. Also, it is important that organizations take a larger cross-functional system view of the assessment and create a template for their overall readiness rather than individual departments producing separate silos of assessments without appreciating the synergies and conflicts within the organization. Millions of dollars are being spent worldwide on managing the risks associated with weather (Lazo et al. 2011) and this assessment tool should help utilities decide whether it is spent on the highest priority ones and aligned with the larger business outcomes.

Next Steps in the Dialogue Between Energy and Meteorology

It is obvious that weather and climate have become more and more important for the energy sector as a) climate change demands an urgent need for adaptation of energy systems on the long term and b) the necessary, increasing share of variable renewables generation—mainly wind and solar—requires quick and significant improvements in weather forecasts, now, on short-term lead times.

As demonstrated in the other chapters in this book, and in many other publications (e.g. Troccoli et al. 2014), the meteorology community has, in the last decade or so, improved knowledge on many energy-relevant aspects; new forecasting tools, methods and products have been made available, and the research and development agenda promises significant new progress in the coming years. But scientific and technical progress alone is not enough to improve energy systems. Indeed, two key additional ingredients are necessary to improve weather readiness (Lazo 2007; Rogers et al. 2007; Dubus 2014):

1. Improved communication between providers and users of the weather and climate information, and
2. Improved decision-making processes.

These two critical aspects will be investigated further in the following chapters (Chaps. 5, 6, 7, 8, 9, 10, 11 and 12) and summarized in the concluding chapter (Chap. 13).

Acknowledgements The authors thank EDF colleagues, especially Marc Trottignon and Vera Silva, for providing material and feedback, and ICEM 2015 energy seminar lecturers: Christian Brose (EON), Mark Zagar (Vestas), Mathieu Anderhalt (EDF) and Marion Schroedter-Homscheidt (DLR).

Appendix: Key Documentation on the Energy Sector

Different organizations gather, analyse, organize and disseminate information about energy worldwide. We list here some of the most relevant ones, and provide the links to the latest releases of their key documents.

- **The World Energy Council** (http://www.worldenergy.org) provides several key publications:

  - World Energy Focus http://worldenergyfocus.org/annual-2016/ http://worldenergyfocus.org/annual-2016/
  - World Energy Resources http://www.worldenergy.org/publications/2016/world-energy-resources-2016/http://www.worldenergy.org/publications/2016/world-energy-resources-2016/
  - Energy Trilemma Index http://www.worldenergy.org/publications/2016/2016-energy-trilemma-index-benchmarking-the-sustainability-of-national-energy-systems/http://www.worldenergy.org/publications/2016/2016-energy-trilemma-index-benchmarking-the-sustainability-of-national-energy-systems/

Many other reports are available on the publications page of the website (perspectives, scenarios …). A nice 4 minute movie summarizes the current status and main challenges in achieving the Energy Trilemma.

- **The International Energy Agency** publishes reference documents every year in November

  - the World Energy Outlook (http://www.iea.org/newsroom/news/2016/november/world-energy-outlook-2016.html). A movie is also available with the 2016 edition. The IEA also produces special reports
  - World Energy Statistics http://www.iea.org/bookshop/723-World_Energy_Statistics_2016http://www.iea.org/bookshop/723-World_Energy_Statistics_2016

Note that IEA documents have an associated cost, but executive summaries can be downloaded freely.

- **The Renewable Energy Policy Network for the 21st Century (REN21)** publishes a Yearly Renewables Global Status Report http://www.ren21.net/status-of-renewables/global-status-report/

Notes

1. http://www.se4all.org/.
2. https://www.worldenergy.org/.
3. https://www.iea.org/.
4. http://www.ren21.net/.
5. It has to be noted that available statistics are generally delayed by 1–2 years, the necessary time for organizations to gather, clean and analyse data from multiple sources.
6. https://www.hydropower.org/.
7. https://www.esmap.org/.

References

Anderhalt, M. (2015). *Power plants and climate change. How to optimize environmental performances of power plants' cooling systems?* ICEM 2015 Energy Seminar, Boulder, Co, USA, June 2015. Retrieved from http://www.wemcouncil.org/wp/wp-content/uploads/2015/07/1415_MathieuAnderhalt.pdf; http://www.wemcouncil.org/wp/wp-content/uploads/2015/07/1415_MathieuAnderhalt.pdf

Baggini, A. (2008). *Handbook of power quality.* Chichester: Wiley, 642pp. ISBN: 978-0-470-06561-7.

Dubus, L. (2014). Weather and climate and the power sector: Needs, recent developments and challenges. In A. Troccoli, L. Dubus, & S. E. Haupt (Ed.), Weather matters for energy (XVII, 528 p.). New York: Springer. ISBN 978-1-4614-9220-7.

Elias, T., Ramon, D., Dubus, L., Bourdil, C., Cuevas-Agullo, E., Zaidouni, T., et al. (2015). Aerosols attenuating the solar radiation collected by Solar Tower Plants: The horizontal pathway at surface level. *AIP Conference Proceedings, 1734*(1). https://doi.org/10.1063/1.4949236.

Elias, T., Ramon, D., Garnero, M. A., Dubus, L., & Bourdil, C. (2016). *Solar energy incident at the receiver of the solar tower plant derived from remote sensing. Part 1: Computation of both DNI and slant path transmittance.* Poster presented at SolarPaces, Abu-Dabi. 10.13140/RG.2.2.19957.37609.

Francis, J., Vavrus, S., & Tang, Q. (2014). Rapid arctic warming and mid-latitude weather patterns: Are they connected? State of the Climate in 2013. Retrieved from http://journals.ametsoc.org/doi/pdf/10.1175/2014BAMSStateofthe Climate.1

Garnero, M. A., Bourdil, C., Dubus, L., Elias, T., & Ramon, D.. (2016). Solar energy incident at the receiver of the solar tower plant derived from remote sensing. Part 2: Impact of the variability of attenuation on plant energy performances. SolarPaces, Abu-Dabi.

HRW. (2014). *Future climate for Africa – scoping papers: Examining the utility and use of long-term climate information for hydropower schemes in sub-Saharan Africa*. Wallingford, UK: HR Wallingford, 41 pp. Retrieved from https://assets.publishing.service.gov.uk/media/57a089b7e5274a27b2000217/hydropower.pdf

IDC. (2013). *Business strategy: Facing down extreme weather*. IDC Energy Insights, #EI242104

IEA. (2016a). *Key world energy statistics 2016*. © OECD/IEA, IEA Publishing. Licence: www.iea.org/t&c. Retrieved from https://www.iea.org/publications/freepublications/publication/KeyWorld2016.pdf

IEA. (2016b). *World energy outlook 2016*. © OECD/IEA, IEA Publishing. Licence: www.iea.org/t&c. Retrieved from http://www.worldenergyoutlook.org/publications/weo-2016/

IHA. (2016). *Hydropower status report, 2016*. International Hydropower Association. Retrieved from https://www.hydropower.org/2016-hydropower-status-report

IPCC. (2013). *Climate change 2013: The physical science basis. Contribution of working group I to the fifth assessment report of the intergovernmental panel on climate change* [Stocker, T. F., Qin, D., Plattner, G.-K., Tignor, M., Allen, S. K., Boschung, J., Nauels, A., Xia, Y., Bex, V., & Midgley, P. M. (eds.)]. Cambridge, UK; New York: Cambridge University Press, 1535 pp. https://doi.org/10.1017/CBO9781107415324. Retrieved from https://www.hydropower.org/2016-hydropower-status-report

Jackson, R. B., Canadell, J. G., Le Quéré, C., Andrew, R. M., Korsbakken, J. I., Peters, G. P., et al. (2015). Reaching peak emissions. *Nature Climate Change, 6*, 7–10. https://doi.org/10.1038/nclimate2892.

LaCommare, K. H., & Eto, J. (2006). Cost of power interruptions to electricity consumers in the United States (U.S.). Lawrence Berkeley National Laboratory, LBNL-58164. Retrieved from https://emp.lbl.gov/sites/all/files/report-lbnl-58164.pdf

Lazo, J. K. (2007). Economics of weather impacts and weather forecasts. In Rose, T. (Ed.), *Elements for life, a WMO publication for the Madrid Conference*. ISBN: 92-63-11021-2.

Lazo, J. K., Lawson, M., Larsen, P. H., & Waldman, D. M. (2011). U.S. economic sensitivity to weather variability. *BAMS*, 709–720. https://doi.org/10.1175/2011BAMS2928.1.

REN21. (2016). *Renewables 2016 global status report*. Paris: REN21 Secretariat. ISBN: 978-3-9818107-0-7.

Rogers, D., Clark, S., Connor, S. J., Dexter, P., Dubus, L., Guddal, J., et al. (2007). Deriving societal and economic benefits from meteorological and hydrological services. *WMO Bulletin*, 56(1), 15–22.

SE4ALL. (2014). *Sustainable energy for all 2014 annual report*. United Nations. Retrieved from http://www.se4all.org/sites/default/files/l/2015/05/SE4ALL_2014_annual_report_final.pdf

Troccoli, A., Dubus, L., & Haupt, S. E., (eds.). (2014). *Weather matters for energy*. New York: Springer. ISBN: 978-1-4614-9220-7 (Print) 978-1-4614-9221-4 (Online).

WEC. (2016a). *World energy perspectives 2016: The road to resilience – managing the risks of the energy-water-food nexus*. World Energy Council Report. ISBN: 978-0-946121-47-2.

WEC. (2016b). *World energy resources, 2016*. World Energy Council Report. ISBN: 978-0-946121-58-8.

WEC. (2016c). *World energy perspectives 2016*. Executive Summary. ISBN: 978-0-946121-51-9.

WEC. (2016d). *World energy trilemma 2016*. Executive Summary. ISBN: 978-0-946121-49-6.

WEC. (2016e). *World energy scenarios 2016*. The grand transition summary report. ISBN: 978-0-946121-57-1.

Open Access This chapter is distributed under the terms of the Creative Commons Attribution 4.0 International License (http://creativecommons.org/licenses/by/4.0/), which permits use, duplication, adaptation, distribution and reproduction in any medium or format, as long as you give appropriate credit to the original author(s) and the source, a link is provided to the Creative Commons license and any changes made are indicated.

The images or other third party material in this chapter are included in the work's Creative Commons license, unless indicated otherwise in the credit line; if such material is not included in the work's Creative Commons license and the respective action is not permitted by statutory regulation, users will need to obtain permission from the license holder to duplicate, adapt or reproduce the material.

CHAPTER 5

Forging a Dialogue Between the Energy Industry and the Meteorological Community

Alberto Troccoli, Marta Bruno Soares, Laurent Dubus,
Sue Haupt, Mohammed Sadeck Boulahya,
and Stephen Dorling

A. Troccoli (✉)
World Energy & Meteorology Council, c/o University of East
Anglia, Norwich, UK

M.B. Soares
University of Leeds, Leeds, UK

L. Dubus
Electricite de France (EDF), Paris, France

S. Haupt
National Center for Atmospheric Research (NCAR) and WEMC, Boulder,
CO, USA

M.S. Boulahya
WEMC, Toulouse, France

S. Dorling
University of East Anglia (UEA) and WEMC, Norwich, UK

© The Author(s) 2018
A. Troccoli (ed.), *Weather & Climate Services for the Energy Industry*,
https://doi.org/10.1007/978-3-319-68418-5_5

Abstract The interplay between energy and meteorology (based on its broad meaning of weather, water and climate) has been steadily growing. For this relationship to continue flourishing, a formal structure for stakeholders to interact effectively is required. The process of formation of the World Energy & Meteorology Council (WEMC), an organisation aimed at promoting and strengthening such a relationship, is discussed in this chapter. Such a process involves building many diverse relationships, something which has been happening over several years, alongside the adoption of more formal practices such as stakeholder surveys. While the focus of this chapter is clearly on WEMC, this process could be used as a stimulus for analogous activities in the broader energy and meteorology area, specifically those at the national and regional levels, as well as similar activities straddling diverse disciplines, such as those promoted by the Global Framework for Climate Services (GFCS).

Keywords Meteorology • Climate services • Energy • Partnerships • Survey • Capacity building • Communication • Outreach • Stakeholder engagement • Education • Associations

Introduction to the World Energy & Meteorology Council

The World Energy & Meteorology Council (WEMC) is a non-profit organisation devoted to promoting and enhancing the interaction between the energy industry and the weather, climate and broader environmental sciences community. Its primary goal is to support improved sustainability, resilience and efficiency of energy systems under ever-changing weather and climate.

Formally established in 2015 as a Company Limited by Guarantee in the UK, WEMC has taken shape over several years. The initial seeds were sown with the 2008 NATO Advanced Workshop, *Weather/Climate Risk Management for the Energy Sector*, which was attended by nearly 30 very active participants who subsequently produced a report published in an international journal (Troccoli et al. 2010) as well as a book that has attracted the attention of thousands of practitioners (Troccoli 2010) and paved the way for more organised interactions between many stakeholders including but not limited to hydrometeorological science and energy sector communities. This start-up workshop was then followed by the more formal and substantial International Conference on Energy & Meteorology (ICEM) series. Four ICEMs have been successfully held thus far—2011 in Australia, 2013 in France, 2015 in USA and 2017 in Italy—with a fifth one to be held in May 2018 in China, where increasing emphasis will be placed on developing world activities and requirements (plans for subsequent conferences are underway). During this period, the WEMC concept has developed substantially thanks to countless discussions amongst experts, the creation of new connections, the burgeoning of the literature in the area of weather, water and climate services as well as work in related international activities such as the International Energy Agency (IEA) Tasks 36 for wind energy, 46 for solar energy and so on. Moreover, the UN-led Global Framework for Climate Services (GFCS) has officially elected energy as a new (fifth) priority area in 2017 and developed a roadmap for the implementation of climate services for the energy industry (WMO 2017, also Ebinger and Vergara 2011).

All of these activities have pointed to the growing interest in strengthening the relationship between energy and meteorology to ultimately help achieve the goals of sustainability, resilience and efficiency of energy systems. In a practical way, WEMC has been acting as the implementing agent for the ideas and recommendations emerging from the ICEMs and beyond (e.g. the GFCS-Energy Exemplar, the World Economic Forum, the United Nations Framework Convention on Climate Change [UNFCCC] Conference of the Parties [COP] 21–Paris Agreement in December 2015). With an average of 200 attendees at each ICEM, an edited book (Troccoli et al. 2014), two special issues in international journals—one in Solar Energy following ICEM 2011 (Troccoli 2013) and one in MetZet following ICEM 2015 (Troccoli and Schroedter-Homscheidt 2017)—and abundant discussions during ICEMs and at related events (e.g. Troccoli et al. 2013), there is a wide-ranging set of issues that naturally feeds into the WEMC concept and plan of action. The plan of action includes (1) the organisation of a series of institutional workshop presentations, webinars and capacity building activities (including internships); (2) the assessments about the significance of future climate projections on energy resource and their implications for energy system investments; (3) the documentation on meteorological and energy data/metadata quality to assist the energy sector to easily access and make optimal use of these data; (4) the formulation and implementation of projects and programmes. These are just some examples amongst the many topics germane to the energy and meteorology intersection. A more comprehensive list of suggested planned activities is presented below, resulting from the WEMC survey.

Rationale for Creating the Organisation

The major ongoing transformation of energy systems worldwide is highlighting the intimate interplay between energy on the one side and weather, climate and water on the other (Green et al. 2016). Although this connection is self-evident in the case of renewable energy (RE), weather and climate information is also critical to a much wider range of energy industry activities, from managing of energy supply from broader energy sources (e.g. offshore oil operations), to the understanding and estimation of energy demand, to the assessment of meteorological impacts on energy extraction, transportation, transmission and distribution (see also Chap. 2).

Given this context, WEMC seeks to substantially contribute to increasing the productivity, resilience and efficiency of energy systems under the influence of ever changing weather and climate as well as to achieving more affordable and available energy, and thereby, foster sustainable and resilient energy systems.

Aims of the Organisation

WEMC aims to enhance productivity and policy formulation for the energy industry through a close collaboration between the energy sector and the weather, water and climate community and to achieve improved adoption of weather, water, climate and other environmental information by the energy industry towards more efficient, proactive and sustainable risk management practices. WEMC also aims to assist the energy industry in meeting the demand for energy while being mindful that there is a need for reducing harmful impacts on the natural environment, in line with international protocols such as the COP21 Paris Agreement.

These aims are tackled through the identification of top-class expertise in energy and meteorological sciences as well as through mobilisation of resources, with fund raising and facilitation of focussed programmes executed by an appropriate mix of energy industry, private service providers, government and international institutions and scientific organisations. Such activities are typically in line with, and in support of, existing relevant national and international efforts.

Ultimately, WEMC aims to create knowledge, critical thinking, experimental tools and funding platforms at the global level for a highly effective use of meteorological information within the energy sector, particularly through creative public–private–academic partnerships. For instance, WEMC supports the energy industry in more effectively utilising meteorological products and facilitating the integration of more appropriate services in a changing climate. Specifically, one key area where the energy sector can benefit from the interaction with the weather and climate community, and that WEMC is contributing to, is the use of meteorological forecasts for grid integration and related tasks such as dynamic line rating. A number of transmission system operators (TSOs) and market operators are already using weather and/or production forecasts for the integration of renewable electricity into the grid. However, this is particularly the case in countries where either RE penetration is important (typically larger than 10%) or use of RE forecasts is mandated (or both). In countries where RE production is still marginal, use of forecasts is not considered a priority. There are also indications that countries/organisations that have taken early action in RE forecasting, rather than to react to a sudden increase in RE production, integrate RE into the grid in a more managed and effective way. It is critical therefore to inform TSOs about the latest developments in meteorological forecasting and the benefits of using this information. Thus, by producing easily accessible, jargon-free and succinct publications, WEMC aims to assist, as in this specific case, with the adoption of meteorological forecasting tools to improve the integration of RE into the grid, and at the same time help control the cost of electricity for consumers.

Structure of WEMC

WEMC is structured around four programmes:

1. Communications and Services
2. Stakeholder Engagement
3. Research and Technology Transfer
4. Education

These programmes, which represent a mix of activities ranging from communication of technical information (programme 1) to a continuous engagement with an ever-widening base of stakeholders as well as the strengthening of advisory bodies at the strategic (Advisory Board) and the technical (Technical Advisory Group) levels (programme 2); to the assistance in the uptake of meteorological tools for more efficient, resilient and sustainable energy systems (programme 3); to capacity building activities such as the ICEMs (programme 4), have clearly been constructed to target the goals of WEMC. Top level international experts are leading the efforts in each of these programmes, with the assistance of the WEMC secretariat, which although currently small and with limited resources, is in a growing phase. The WEMC secretariat is sited on the campus of the University of East Anglia (Norwich, UK).

Sustained input from international experts, particularly the ICEM Organising Committee and the WEMC Technical Advisory Group, is key to the success of the WEMC initiative, in a similar manner to the way that related international activities function (e.g. the above-mentioned IEA tasks). Indeed, assistance from a wide base of experts is essential to tackle prioritised activities aimed at improving the interaction between energy and meteorology (e.g. data exchange and their standardisation).

WEMC heavily relies on the expert guidance of its Advisory Board, which is drawn from as diverse a group of people as possible to cover the main international players in the energy sector and the meteorology community, both in the developed and developing countries and including related associations (e.g. the USA-based Utility Variable Integration Group, UVIG).

Also integral to the structure of WEMC are Special Interest Groups (SIGs). The growing WEMC membership is leading the work of the SIGs. The plan is to constitute several such SIGs, one for each identified main critical activity, with around a dozen experts participating in each. Initially WEMC will focus on three initial SIGs, one on Data Sharing and Standards, the second on Grid Integration and the third on Education. This choice has been informed by members and the interests of other experts' interest, as identified through the survey presented in the next section. WEMC will provide communication tools and assistance to facilitate the work of these groups. It will also horizon scan for funding opportunities and pursue the most relevant and promising ones so as to provide financial support to the activities of specific working groups (e.g. to hold physical meetings). These SIGs are tasked to produce output readily usable by the energy industry such as the aforementioned guidance document on grid integration (Fig. 5.1).

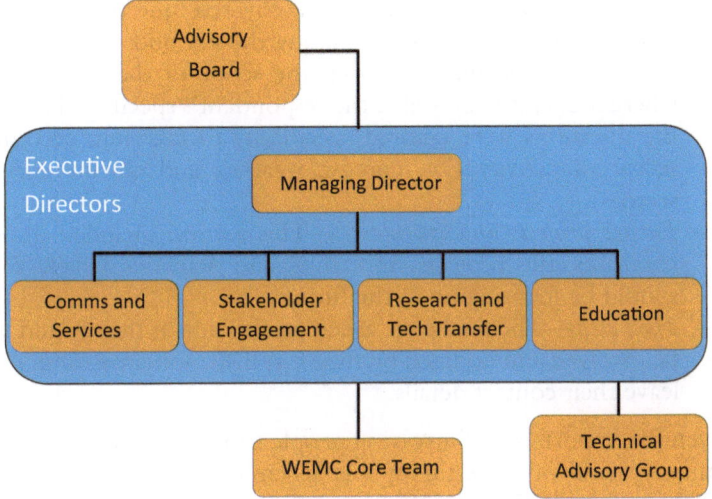

Fig. 5.1 The World Energy & Meteorology Council (WEMC) organigram

Defining Priorities for WEMC: The Users' Survey

Rationale for Undertaking a Survey

Given the aim and nature of the WEMC activities, it was fundamental to define the next steps for the organisation based on the needs and requirements of the potential future users of those services. A survey was conducted with the target of engaging with as many potential users and organisations around the world as possible.

Methodology and Implementation of the WEMC Survey

The method selected to engage with potential users of WEMC was an online survey as it enabled the collation of the widest possible number of responses from around the world as possible in a relatively short amount of time (May 2011). The survey, developed using the software Survey Monkey, included four main sections[1]:

- *Welcoming page*—This introductory page described WEMC and asked respondents to select their sector of activity: energy; weather and climate; or other.

- *Your organisation*—This section covered questions about the respondent's organisation including the size and type of organisation, its geographical location, the scope of their activities and where it operates, as well as the respondent's specific role. Some of the questions were tailored specifically to take into account the sector of activity (i.e. energy; weather and climate; or other sectors).
- *Future projects and initiatives*—This section included questions regarding the projects, initiatives and activities which WEMC should be focusing on in the future.
- *Next steps*—The final page asked participants if they would like to be involved and updated on future WEMC initiatives and, if so, to leave their contact details.

Given the importance of involving people from the energy and weather community worldwide, the survey was disseminated to participants at the 3rd ICEM in Boulder USA in 2015[2] as well as circulated to targeted mailing lists such as the Energy-L[3] and Climate-L,[4] mailing lists for energy policy issues and climate-related news, respectively. The survey was officially launched in June 2015 and closed in January 2016. The sections below describe some of the main findings from the survey and how it helped inform future activities within WEMC.

Results from the WEMC Survey

A total of 147 responses were received between June 2015 and January 2016. Almost half of the respondents worked in the energy sector (47%, $n = 69$), followed by those working on weather and climate related activities (33%, $n = 39$) and other sectors (20%, $n = 29$). The type of organisations also varied with private companies representing the large majority in the energy sector whilst research institutes were the most represented in the weather and climate sectors (Fig. 5.2).

The participants' organisations also varied in size and between the sectors of analysis. The energy sector showed the highest number of large companies (with more than 5000 employees) followed by smaller organisations (with up to 100 employees). Conversely, the weather and climate sector showed the highest number of smaller companies (up to 100 employees) followed by those with between 1000 and 5000 employees (Fig. 5.3). Organisations in the 'other' sector had a fairly similar distribution

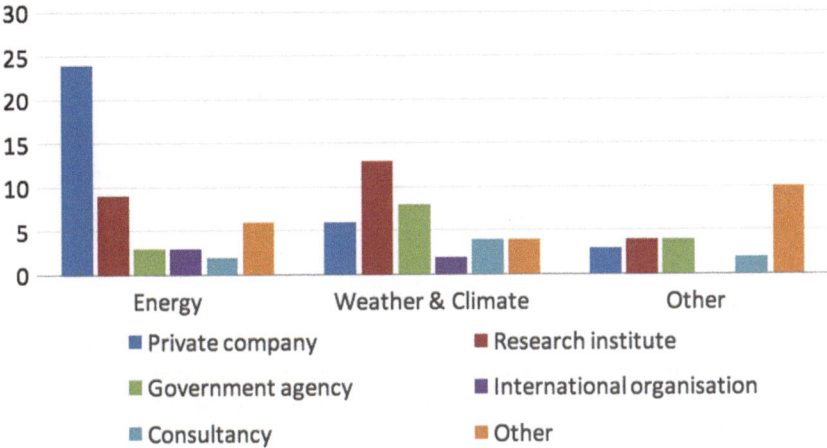

Fig. 5.2   Type of organisations per sector of activity

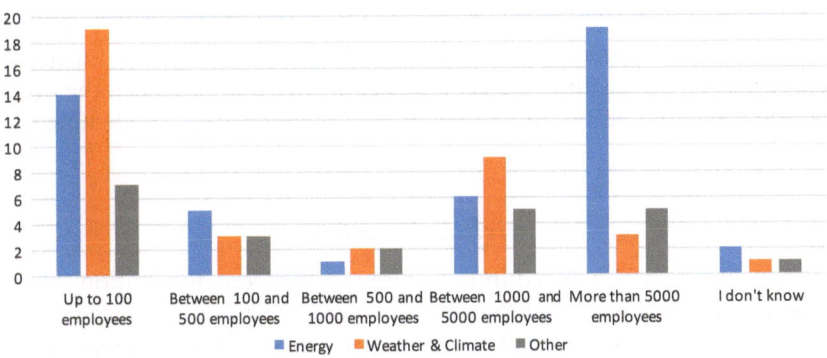

Fig. 5.3   Size of the responding organisations per sector

with regard to the size of the companies who responded to the survey (Fig. 5.3).

The respondents from the energy sector were mainly based in Europe (France, Germany, Denmark, Spain) as well as the USA; whilst those working in the weather and climate sector were mostly based in the USA followed by France and the UK (Fig. 5.4).

With regard to the scope of the organisations' activities, these were predominantly worldwide for both the energy and the weather and climate

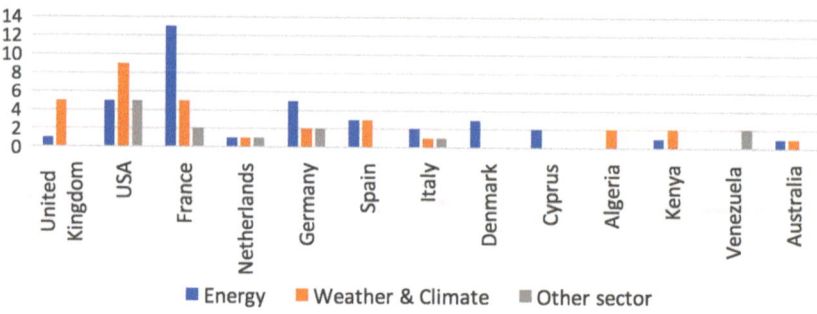

Fig. 5.4 Countries where survey respondents are based. Countries selected by only one respondent were excluded from this chart. These included Brazil, Austria, Vietnam, Costa Rica, Namibia, South Africa, New Zealand, Mexico, Zambia, Greece, Indonesia, Argentina, Malaysia, Bosnia and Herzegovina, India, Finland and Guatemala, Ghana, Morocco, Chad, United Arab Emirates and Mauritania

sector followed by those operating across Europe and in specific countries worldwide for the energy sector and those operating in specific European countries for the weather and climate sector (Fig. 5.5). Those in the other sectors operated across the range of geographical areas as identified in Fig. 5.5.

Activities Across Sectors

The main activities pursued in the energy sector were distribution/transmission, technology development, and power development (Fig. 5.6), although approximately half of the organisations in the energy sector ($n = 25$) worked in two or more activities.

Similarly, more than half of the organisations ($n = 27$) operated in two or more areas within the renewables (i.e. solar, wind and hydroelectric power) which was the sub-sector most strongly represented amongst the surveyed organisations (Fig. 5.7). Given renewables are, amongst the energy systems, the most impacted by weather and climate events, this result was not unexpected; however, this also reflects the backgrounds and interests of the respondents.

Another interesting aspect was the fact that 75% of these organisations ($n = 35$) were involved in the energy and meteorology nexus (i.e. working in areas linking energy and meteorology). The organisations working with

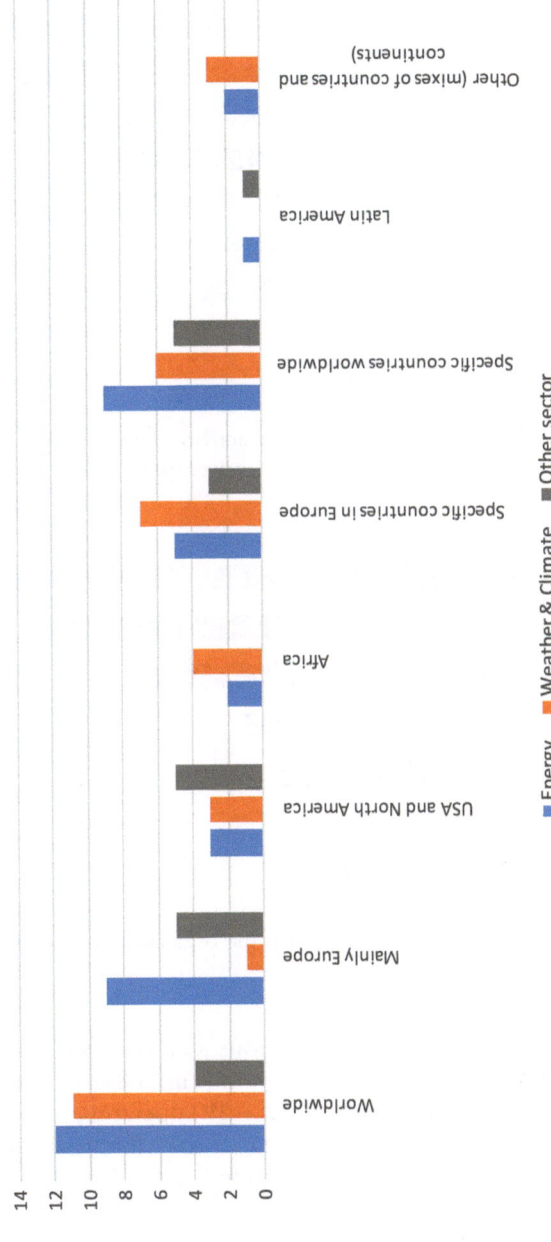

Fig. 5.5   Organisations scope of operations and activities

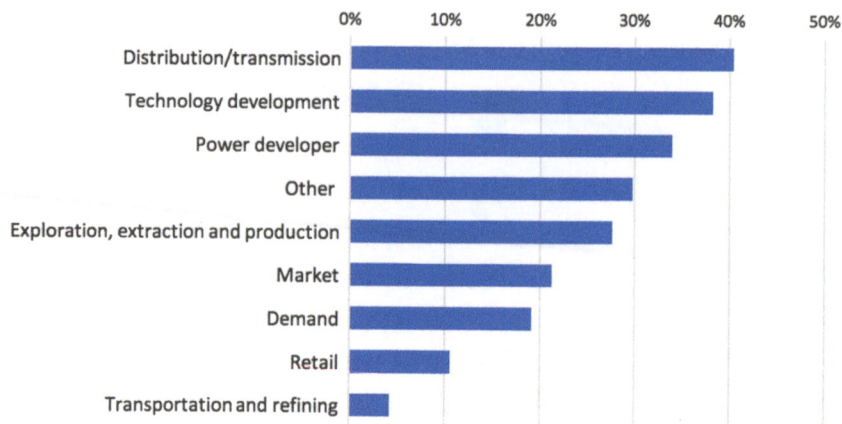

Fig. 5.6 Scope of responding organisations' activities in the energy sector (total per cent of $n = 47$; note that this was a multi-answer question)

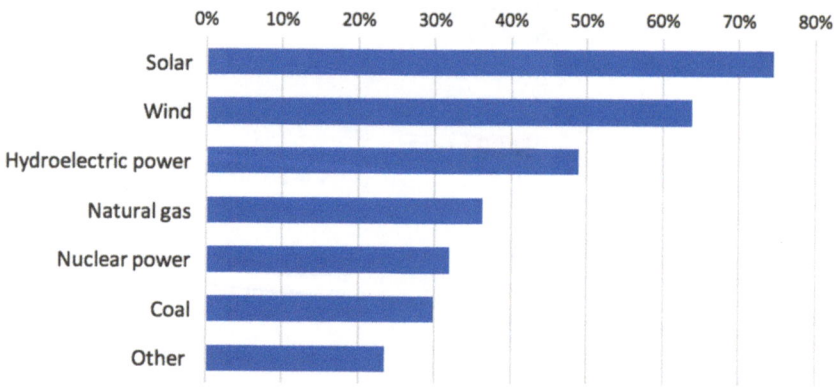

Fig. 5.7 Area of the energy sector in which the organisations operate (note that this was a multi-answer question)

weather and climate were fairly homogenous with regard to the provision of different weather and climate information (Fig. 5.8).

The remainder of the organisations ($n = 29$) operating in other sectors was mainly constituted by those working in academic research, government and public administration, forestry, media, biodiversity and ecology and coastal activities.

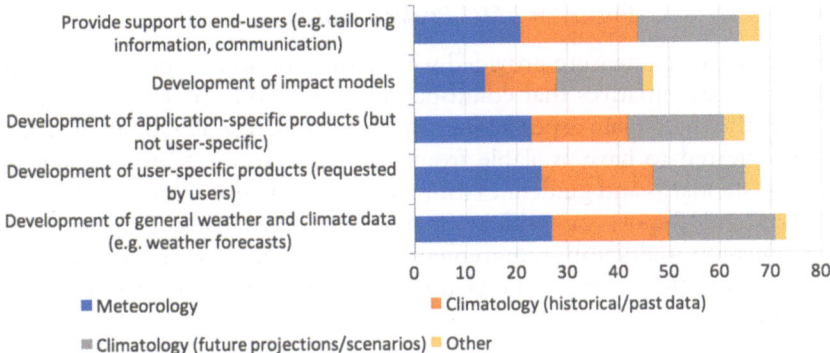

Fig. 5.8 Scope of activities in the responding organisations operating in meteorology and climate (note that this was a multi-answer question)

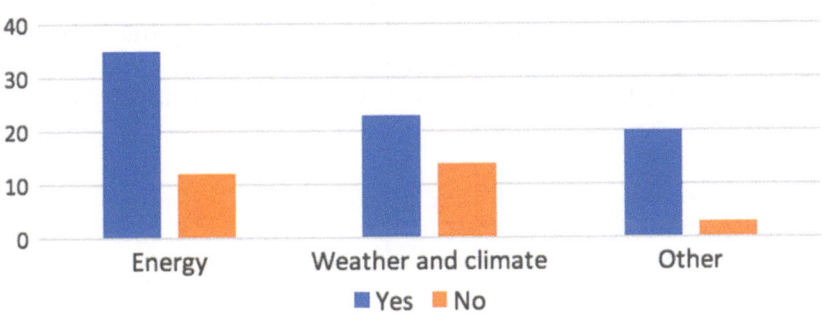

Fig. 5.9 Number of organisations surveyed interested in the energy and meteorology nexus

Nexus Between Energy and Meteorology

Survey participants were also asked about the interest of their organisation in the energy and meteorology nexus. The large majority of respondents confirmed the interest in this nexus with approximately 75%, 62% and 87% of the respondents in the energy, weather and climate and other sectors agreeing, respectively (Fig. 5.9).

Future WEMC Projects and Initiatives

Survey participants were provided with a list of options of potential policy and services initiatives that could be pursued by WEMC (Fig. 5.10). Of those, the three main aspects that respondents (across all sectors of analysis) preferred to have available from WEMC were the 'Development of codes, standards and guidelines for meteorological information'; 'Position papers' and 'Reports on resilience and sustainable energy systems and links to emission reduction requirements'. Conversely, the least preferred option was 'Recommendations on data/metadata quality to assist the energy sector' (Fig. 5.10).

It should be noted, however, that the range of scores between the most and least popular suggested policy/services initiatives is relatively small, and their differences are likely within the sample error. In addition, given this is an evolving area we expect these responses to vary over time. Therefore, it is not straightforward to clearly pinpoint which activities are deemed as critical to the extent that they should be prioritised.

Participants were also asked about their preferences regarding research and technology transfer initiatives that should be pursued by WEMC. The 'Development of methodologies for analysing the linkages between

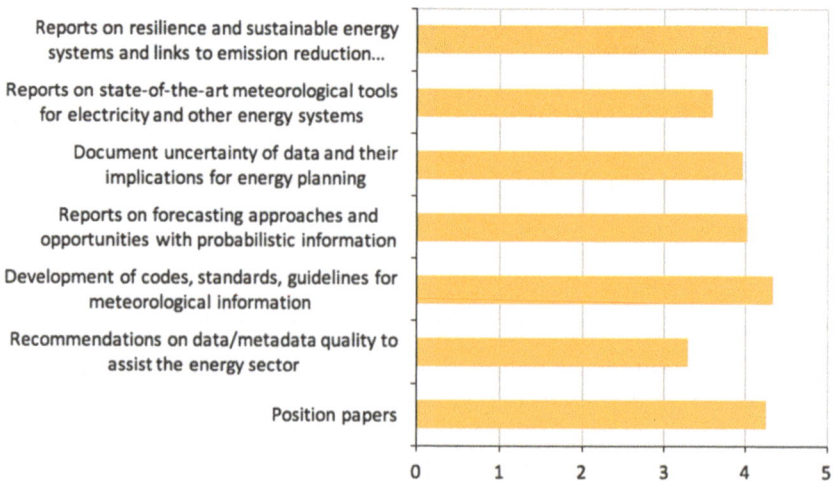

Fig. 5.10 Preferences from survey respondents regarding policy/services initiatives to be pursued by WEMC (based on rating average of ranked preferences)

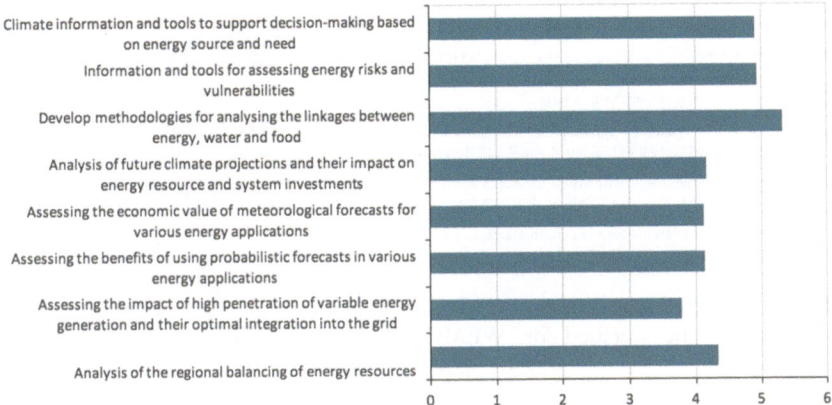

Fig. 5.11 Preferences from survey respondents regarding research and technology transfer initiatives to be pursued by WEMC (based on rating average of ranked preferences)

energy, water and food' ranked the highest of all the options given (Fig. 5.11). This was followed by 'Information and tools for assessing energy risks and vulnerabilities' and 'Climate information and tools to support decision-making based on energy source and need'. The least preferred option was 'Assessing the impact of high penetration of variable energy generation and their optimal integration into the grid' (Fig. 5.11).

Finally, participants were asked about their preferences regarding outreach and training activities. The main priority for respondents is the creation of a 'WEMC mailing list and newsletter' (Fig. 5.12). Following from that, respondents were also interested in 'Targeted schools on particular topics within the energy and meteorology nexus' and 'A series of online webinars on the energy and meteorology nexus'. The least preferred option was 'Creation of a database of organisations, projects, events, and best practices to support potential collaborations' (Fig. 5.12).

Paying for WEMC Services

Participants were also asked about their willingness to pay for WEMC services. Of those who responded to this question ($n = 88$), 47% agreed that they would be willing to pay for those services whilst 53% disagreed. Those who indicated they would be willing to pay were then asked about

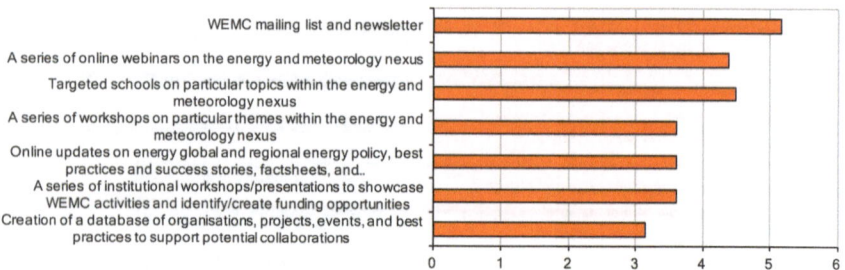

Fig. 5.12 Preferences from survey respondents regarding outreach and training activities to be pursued by WEMC (based on rating average of ranked preferences)

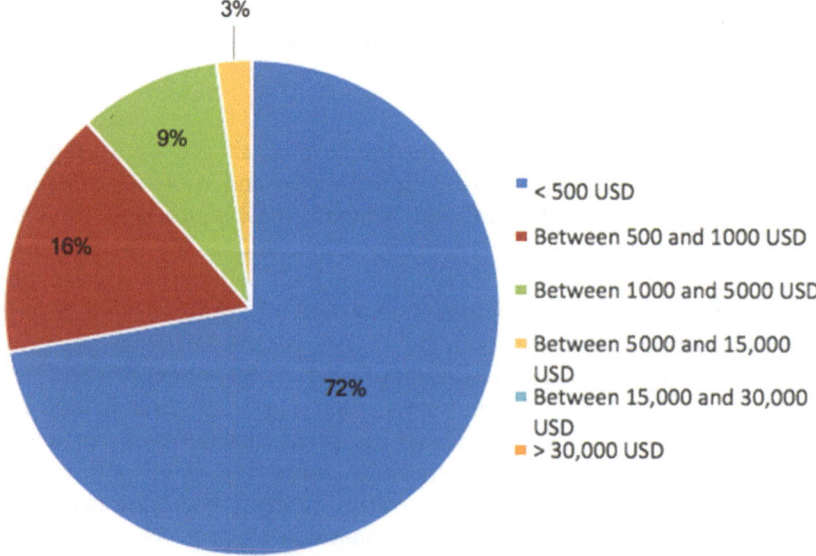

Fig. 5.13 Respondents' willingness to pay for WEMC services, per year

how much they would pay per year for the WEMC services. Around 72% of the respondents would be willing to pay less than 500 US dollars per year, whilst only 16% would pay between 500 and 1000 US dollars, 9% between 1000 and 5000 US dollars and only 3% said they would be willing to pay between 5000 and 15,000 US dollars (Fig. 5.13).

Of the total number of respondents, 46% agreed to continue being involved and updated regarding future WEMC activities (including those both willing and not willing to pay for the services). These mainly included respondents from Europe (56%) followed by those from the USA (21%), Africa (14%), South America and the Caribbean (6%) and Asia (3%).

Next Steps for WEMC

The results of the survey have been very useful in shaping and guiding WEMC activities thus far, and for the immediate future. Specifically, the results of the survey have been useful to highlight:

(a) The interest of survey respondents in WEMC pursuing a number of activities such as 'Development of codes, standards and guidelines for meteorological information' or 'Development of methodologies for analysing the linkages between energy, water and food';

(b) The indication from nearly 50% of the respondents of their interest to remain engaged in and informed about future activities in the energy and meteorology nexus;

(c) The willingness of a comparable percentage of respondents to pay for services provided by WEMC, a clear indication that these services are deemed important and valuable.

It is also important to note that the results of the survey were written up nearly a year after it was closed. Having had this additional period to see the evolution of the sector, particularly in terms of stakeholders' requirements, it is apparent that some of the priorities have somewhat shifted since then. For instance, the optimal integration of RE into the grid was given as a low priority by the respondents (Fig. 5.11), while there is evidence (as highlighted earlier in this chapter and also through the work of, e.g., the UVIG[5]) that this is now higher in experts' agendas. This apparent shift may also be indicative of the fact that our survey sample, although reasonably large, was not robust enough to clearly discriminate amongst priorities areas. It may also simply be a reflection of the fact that preferences regarding policy/services initiatives (Fig. 5.10) cannot capture the nuances in the response choices available. In all, stakeholder consultations, through surveys similar to the one presented here or via other processes (e.g. workshops), will need to be an integral component of WEMC activities so that new ideas, needs and other information from stakeholders are taken into consideration in a timely fashion.

Partnerships with analogous organisations (e.g. the International Solar Energy Society, ISES) are also key to the success of WEMC, and these are being actively pursued, for instance, via the co-organisation of webinars, a communication tool that is proving very popular or the involvement of key people (e.g. ISES president) in the WEMC Advisory Board.

Another important aspect of the WEMC activities is to provide a blueprint, and ideally an international reference, for national and regional activities in the area of energy and meteorology. A few such activities have already been initiated, with the USA (e.g. American Meteorological Society's annual Conference on Weather, Climate, Water and the New Energy Economy, which started in 2009) and the EU (the Energy Meteorology session at the European Meteorological Society annual conference, which also started in 2009) leading the way. More recently, a meeting on energy and meteorology was held in China in 2016 for the first time, for which ICEM was taken as a reference for the organisation of the event (Dr Rong Zhu, China Meteorological Administration, personal communication).

Overall, what is clear is that a continuous, adaptable and proactive interaction will be required in order to make WEMC's activities valuable to a wide range of stakeholders in this relatively fast evolving interdisciplinary area. Further discussion about next steps in the area of energy and meteorology is presented in the final chapter of this book.

NOTES

1. The survey questions are available at: http://www.wemcouncil.org/MEMBERS/WEMC_Survey_Qs_2015.pdf.
2. http://www.wemcouncil.org/wp/conferences/icem2015/.
3. https://lists.iisd.ca/read/?forum=energy-l.
4. https://lists.iisd.ca/read/?forum=climate-l.
5. https://www.uvig.org/.

REFERENCES

Ebinger, J., & Vergara, W. (eds.). (2011). *Climate impacts on energy systems: Key issues for energy sector adaptation*. World Bank Publication. Retrieved from http://www.esmap.org/sites/esmap.org/files/E-Book_Climate%20Impacts%20on%20Energy%20Systems_BOOK_resized.pdf

Green, J. M. H., Cranston, G. R., Sutherland, W. J., et al. (2016). Research priorities for managing the impacts and dependencies of business upon food, energy, water and the environment. *Sustainability Science.* https://doi.org/10.1007/s11625-016-0402-4.

May, T. (2011). *Social research.* Maidenhead, UK: McGraw-Hill Education.

Troccoli, A. (ed.). (2010). *Management of weather and climate risk in the energy industry.* NATO Science Series. Dordrecht: Springer Academic Publisher. Retrieved from http://www.springer.com/earth+sciences+and+geography/atmospheric+sciences/book/978-90-481-3690-2

Troccoli, A. (2013). ICEM Solar Radiation, Solar Energy, 98 Part B, 99. https://doi.org/10.1016/j.solener.2013.10.030

Troccoli, A., & Schroedter-Homscheidt, M. (2017). Special issue on the 3rd International Conference on Energy Meteorology, 22–26 June 2015. *Meteorologische Zeitschrift, 26*(3), 237. Retrieved from https://www.schweizerbart.de/content/papers/download/87596.

Troccoli, A., Boulahya, M. S., Dutton, J. A., Furlow, J., Gurney, R. J., & Harrison, M. (2010). Weather and climate risk management in the energy sector. *Bulletin of the American Meteorological Society, 6,* 785–788. https://doi.org/10.1175/2010Bams2849.1.

Troccoli, A., Audinet, P., Bonelli, P., Boulahya, M. S., Buontempo, C., Coppin, P., et al. (2013). *Promoting new links between energy and meteorology. Bulletin of the American Meteorological Society,* ES36–ES40. https://doi.org/10.1175/BAMS-D-12-00061.1.

Troccoli, A., Dubus, L., & Haupt, S. E. (eds.). (2014). *Weather matters for energy.* New York: Springer Academic Publisher, 528 pp. Retrieved from http://www.springer.com/environment/global+change+-+climate+change/book/978-1-4614-9220-7

WMO. (2017). *Energy exemplar to the user interface platform of the global framework for climate services.* World Meteorological Organisation, 120 pp. Retrieved from https://library.wmo.int/opac/doc_num.php?explnum_id=3581

Open Access This chapter is distributed under the terms of the Creative Commons Attribution 4.0 International License (http://creativecommons.org/licenses/by/4.0/), which permits use, duplication, adaptation, distribution and reproduction in any medium or format, as long as you give appropriate credit to the original author(s) and the source, a link is provided to the Creative Commons license and any changes made are indicated.

The images or other third party material in this chapter are included in the work's Creative Commons license, unless indicated otherwise in the credit line; if such material is not included in the work's Creative Commons license and the respective action is not permitted by statutory regulation, users will need to obtain permission from the license holder to duplicate, adapt or reproduce the material.

CHAPTER 6

Weather, Climate and the Nature of Predictability

David J. Brayshaw

Abstract The prediction and simulation of future weather and climate is a key ingredient in good weather risk management. This chapter briefly reviews the nature and underlying sources of predictability on timescales from hours-ahead to centuries-ahead. The traditional distinction between 'weather' and 'climate' predictions is described, and the role of recent scientific developments in driving a convergence of these two classic problems is highlighted. The chapter concludes by outlining and comparing the two main strategies used for creating weather and climate predictions, and discussing the challenges of using predictions in quantitative applications.

Keywords Weather prediction • Climate prediction • Predictability • Chaos • Modelling

INTRODUCTION

A long-standing challenge for meteorology and climate science has been to develop techniques capable of producing predictions and simulations of the weather and climate across a range of timescales. Although these

D.J. Brayshaw (✉)
Department of Meteorology, University of Reading, Reading, UK

National Centre for Atmospheric Science, Reading, UK

© The Author(s) 2018   85
A. Troccoli (ed.), *Weather & Climate Services for the Energy Industry*,
https://doi.org/10.1007/978-3-319-68418-5_6

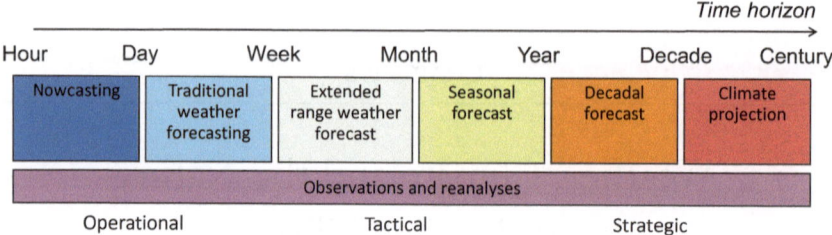

Fig. 6.1 Weather and climate timescales, forecasting tools and datasets

timescales should be viewed as a continuum, it is nevertheless helpful to identify a number of discrete timescales, ranging from very near term 'nowcasting' to climate change projections over centuries and millennia, as shown in Fig. 6.1.

This chapter briefly reviews how the nature of predictability differs across prediction timescales, identifies the major strategies used to create predictions and discusses some of the general challenges with using predictions in quantitative applications.

THE NATURE OF PREDICTABILITY

To understand weather and climate forecasting, it is important to understand the character of the physical system one is seeking to predict. The atmosphere (and the climate system more broadly) can be viewed as an example of a chaotic system (e.g., Lorenz 1963), associated with two distinct 'types' of predictability (Lorenz 1975, see also Schneider and Griffies 1999). These types—and their relationships to different timescales of weather and climate—are discussed herein.

Due to the complexity of the atmosphere (or climate) system, it is helpful to discuss predictability with reference to an analogous but simpler chaotic system. The Lorenz model contains three inter-dependent variables (U, X and Y) evolving deterministically over time. Each of the three dimensions can be understood as representing a meteorological quantity in analogy (e.g., eastward wind, northward wind and temperature). A typical atmosphere-only climate model will, however, have in excess of ~10^6 dimensions: one each for six key meteorological properties (wind in the horizontal and vertical, temperature, water vapour and surface pressure) at each point on a 3-dimensional grid (perhaps $192 \times 120 \times 30$).

In some cases, the time evolution of the equations may also include a stochastic (random) component rather than being purely deterministic. The equations of the Lorenz model may be written in finite difference form:

$$X_t = X_{t-1} + \left[\alpha\left(Y_{t-1} - X_{t-1}\right)\right]\delta t$$

$$Y_t = Y_{t-1} + \left[X_{t-1}\left(\beta - U_{t-1}\right) - Y_{t-1} + \epsilon\right]\delta t$$

$$U_t = U_{t-1} + \left[X_{t-1}Y_{t-1} - \gamma U_{t-1}\right]\delta t$$

where α, β and γ are constant parameters, subscripts denote time-steps, and δt is the interval between adjacent time-steps. Following Palmer (1999), ϵ is used to denote a small external forcing (for the initial discussion it is assumed that $\epsilon = 0$).

The time evolution of the Lorenz system can be represented as a trajectory (or path) in phase space.[1] Figure 6.2a shows a short section of a trajectory as an example: from an initial state near $(U, X, Y) = (33, 15, 18)$, the model evolves to a state $(24,-12,-18)$ over a 'time' interval $\sum \delta t = 0.7$. If the model is allowed to evolve for a longer period to produce a more extended trajectory (referred to as an attractor), a fuller view of the system's properties emerges (Fig. 6.2b). The attractor clearly shows two lobes, with the system preferring to occupy states in one or the other of the lobes.

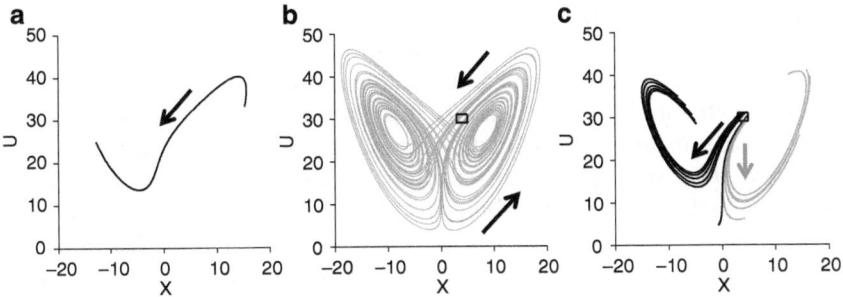

Fig. 6.2   The Lorenz model and initial condition problems, using $\alpha = 10$, $\beta = 28$, $\gamma = 8/3$ and $\epsilon = 0$. See text for discussion

Figure 6.2 can be used to illustrate an example of an *initial condition problem* or 'predictability of the first kind'. Consider at some time t_1 that observations of the system are taken, and found to be $U_1 = 30$ and $X_1 = 4$. Assume that each of these measurements is subject to an observational error, $\Delta U_1 = \Delta X_1 = 1$, and Y_1 is unobservable in practice.[2] The systems current position in phase space is, therefore, not known exactly, but can be constrained to a relatively small region of phase space indicated by the black box in Fig. 6.2b. A prediction of the system's state at some future snapshot in time, $t_2 = t_1 + \Delta t$ is sought—that is, we wish to predict the exact values of U_2 and X_2. This is analogous to weather forecasting: we 'know' the weather today and wish to predict the weather tomorrow.

There are many possible phase space trajectories that are consistent with the available knowledge of the initial conditions at time t_1. A selection of these are shown in Fig. 6.2c: some lead to the left lobe (black lines), whereas others remain in the right lobe B (grey lines). In consequence, a relatively small error in estimating the starting state ($\Delta U_1 = \Delta X_1 = 1$) grows rapidly to a large error in the prediction ($\Delta U_2 \sim \Delta X_2 \sim 30$). The rate of error growth is, however, very dependent on the initial state and the forecast time horizon considered; in this example there is very low predictability but, if the initial conditions correspond to some other regions of the attractor, there may be much more predictability (i.e., smaller errors), at least over short time horizons (i.e., small Δt).

The evolution, shape and position of a trajectory are also sensitive to the model's parameters (α, β and γ), typically referred to as boundary conditions. Figure 6.3 shows the original attractor from the previous figure (in black) and a new attractor (in grey)—the only difference is a small change in one of the boundary conditions, ϵ. Clearly, one can detect a change in the probability distribution of the observable quantity, U, as indicated by the relative frequency distributions in the bottom panel in Fig. 6.3. This is an example of 'predictability of the second kind', which concerns the ability to predict changes in the attractor in response to changes in *external boundary conditions*.[3] Clearly, if the response is large, then it can be more readily detected against the 'internal' variability corresponding to trajectories moving within a single attractor.

Traditional climate change simulations studying the equilibrium climate under a future greenhouse gas concentration scenario can be viewed

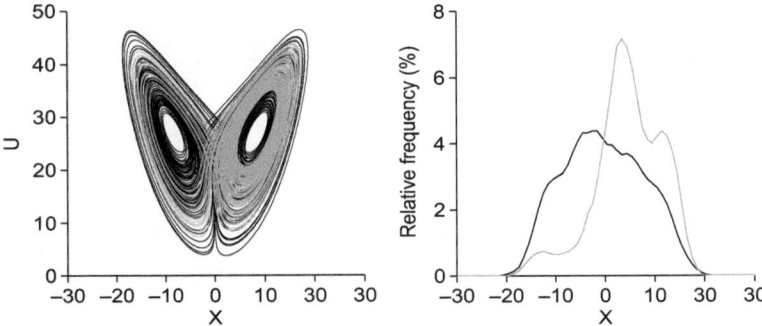

Fig. 6.3   The Lorenz model and the long-term equilibrium climate change problem. The black and grey curves show two simulations with different boundary conditions (parameters as in Fig. 6.2, but with $\epsilon = 10$ for the grey curve). See text for discussion

as an example of second kind predictability: it is a boundary condition prediction problem where one seeks to understand how the statistics of climate differ between two different sets of boundary conditions (e.g., Meehl et al. 2007).

If numerical weather prediction (NWP) (days-ahead) and long-term climate change simulations (decades-ahead) can be considered as examples of initial condition and boundary condition problems, then it is clear that much lies between these two extremes. It is therefore helpful to consider the timescales involved in the system one is seeking to predict.

The climate system in general contains many different components, varying on a wide range of timescales (Fig. 6.4). At forecast lead times of 1–2 days, it is typically sufficient to focus on the evolution of the faster components alone (e.g., troposphere and land-surface temperature) as the slower components (e.g., ocean temperature, ice sheets) change little during the lifetime of the prediction. Indeed, at very short lead times (minutes to hours) many aspects of the large-scale flow in the troposphere may even be considered fixed. Conversely, at longer forecast lead times, the evolution of slower components become significant (e.g., ocean circulation, land-surface moisture, ice sheets, sea ice and snow cover). For a prediction of tomorrow's weather in London, it may, therefore, be sufficient

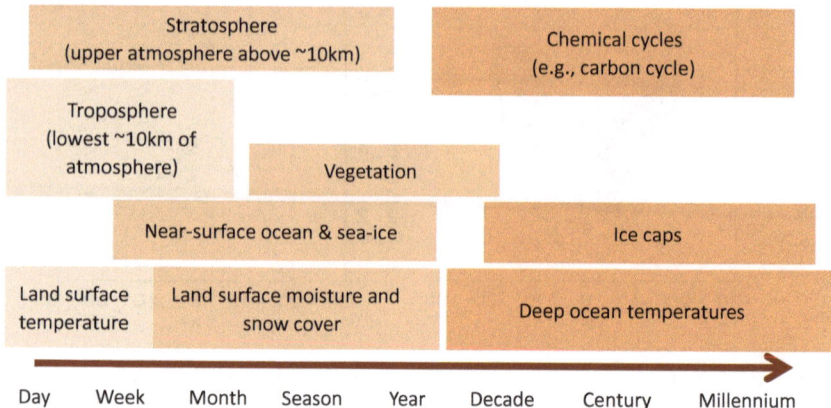

Fig. 6.4   Indicative timescales of selected components in the climate system

to neglect small changes in the temperature of the North Atlantic ocean's surface, but the same cannot be said for predicting the seasonal-average temperature over Europe several months in advance.

Weather and climate predictions (as outlined in Fig. 6.1)—particularly those in the range of several days to a few decades—are therefore a mixture of initial and boundary condition problems, and have been the subject of much research in recent years. This can be illustrated by considering, for example, a seasonal forecast. In such a forecast, the state of the troposphere and land-surface temperature change much more quickly (~days) than the forecast horizon (~months). From the perspective of these components, the problem is therefore boundary condition prediction and predictability of the second kind (i.e., the intention is to predict the statistical properties of the troposphere rather than estimate its state at a specific snapshot in time). However, for the ocean surface, soil moisture, snow cover and stratosphere—which only change slowly over the timescale of the forecast—the challenge is to determine the specific evolution, therefore concerns predictability of the first kind where initial conditions play a key role.

Prediction Strategies

There are two broad categories of predictive models used in weather and climate forecasting: statistical and dynamical. The primary characteristics of each are outlined below.

Statistical Models

Statistical predictions come in many forms and are in widespread use throughout academia and industry for many problems. Conceptually, these models are simple: historical observations are interrogated to find relationships between a predictand and a set of potential predictors. These historical relationships are then assumed to remain fixed into the future and are used to create a prediction. Common examples include single- or multi-variate autoregressive models (or more sophisticated versions such as ARMA and GARCH), artificial neural networks, support vector machines and 'analogue'-based techniques.

Statistical models are an undoubtedly powerful prediction tool. It must, however, be recognised that there are limitations and dangers associated with this approach. The process of statistical modelling is essentially ad-hoc: the predictors to be used are not necessarily known a priori (unless informed by some prior physical or dynamical process understanding), and must, therefore, be established afresh for each new predictand. The ability to identify statistical relationships between predictors and predictand is also constrained by the quantity and quality of the available historic data: the records must be sufficiently long and homogeneous to robustly establish statistical relationships between variables. Finally, statistical models trained on historic data may ultimately be a rather poor guide for a climate system subject to changing boundary conditions (e.g., greenhouse gas concentrations). This is perhaps particularly the case for longer range forecasts where many plausible future states will simply not have been recorded in historic observations

Dynamical Models

In contrast to statistical models, dynamical prediction models numerically simulate the behaviour of the system itself and are the basis of the weather forecasts provided by most operational weather services. In NWP, the atmosphere is represented by fundamental physical equations based on the laws of motion, thermodynamics and conservation of mass. The equations are discretised in space and time (i.e., the atmosphere is divided into grid boxes) and solved iteratively in each grid box, advancing time-step by time-step. Additional physical and dynamical processes such as clouds, precipitation and radiation are represented through 'parameterization schemes' at the grid box level.

As discussed in section 'The Nature of Predictability', initial condition errors in NWP can grow rapidly over a short period (Fig. 6.2c) and 'ensemble forecasting' is widely used. An NWP ensemble consists of a large set of N individual 'member' weather forecast simulations, each starting from a slightly different set of initial or boundary conditions (all of which are consistent with the observations, subject to observational error). Typically, the set of initial conditions chosen will seek to maximise the difference between the ensemble members at the targeted forecast time—that is, to capture the widest possible range of uncertainty associated with errors in the initial conditions. Such forecasts, although still 'predictions of the first kind', provide N different potential realisations of the future weather state and must be interpreted probabilistically (typically $N \sim 50$ in many present NWP systems) with the 'spread' of outcomes in a good forecast system providing an indication of the predictability available from the particular set of initial conditions used.

Beyond a few days to a couple of weeks, additional climate system components must also be included in addition to the atmosphere (see Fig. 6.4 and section 'The Nature of Predictability'), such as the stratosphere, oceans and sea ice. The resulting dynamical models draw strongly on the heritage of General Circulation Models (GCMs). First developed in the 1960s–1980s—see, for example, Smagorinsky et al. 1965 for a very early example—and continually developed since, GCMs are identical in concept to NWP models—insofar as they represent a physical model of the system one is attempting to predict or understand—but as they include more physical processes and must be run over longer timescales, they typically use much coarser resolution grid boxes than NWP. As in NWP, the use of ensembles in climate model simulations is common. In this case, however, the ensemble is typically used to sample several different sources of uncertainty[4]: natural climate variability (Deser et al. 2012), initial condition uncertainty (Scaife et al. 2014), parametric uncertainty (Stainforth et al. 2005) or model structural uncertainty (Taylor et al. 2012).

The power of dynamical models to simulate weather and climate is considerable. The skill of NWP models at lead times of several days ahead has increased continuously over recent decades, and GCMs are now used to produce very sophisticated realisations of physical phenomena affecting a wide range of industrial sectors. There are, however, limitations. NWP and GCM models are computationally expensive compared to statistical models, leading to a three-way trade-off between resolution (grid size), physical complexity (number of processes modelled) and computational feasibility (number of model years simulated and ensemble size). All dynamical models are subject to biases and/or growth of prediction error from many sources, such as deficiencies in model formulation (numerical approximations, missing processes), parametric uncertainty (ill-constrained properties in parameterisation schemes) and initial conditions. Before predictions are used, care should be taken to establish whether dynamical models produce a reliable representation of any particular phenomena of interest

SUMMARY AND DISCUSSION

There are good reasons to believe that predictability exists in weather and climate forecasting across a range of timescales from hours to decades and beyond. This predictability may take one of two forms: either a prediction of the specific evolution of the weather (an initial condition problem) or else a prediction of the statistical properties of the climate (a boundary condition problem). Weather and climate forecasts in the intermediate range (several days to decades) typically incorporate some aspects of both forms of predictability, and a probabilistic approach to the resulting forecast is essential.

Both the statistical and dynamical approaches discussed above have great power in terms of achieving predictive skill. It is, however, emphasised that the two approaches should be seen as being complementary toolkits rather than competing philosophical strategies. Statistical methods are often used to 'calibrate' dynamical model output (reduce bias when compared against point observations) and configure dynamical models forecasts (e.g., by statistically identifying key boundary conditions such as sea-surface temperature patterns). Conversely, dynamical models enable deeper process understanding (helping to identify robust predictors for statistical models) and—with care—can be used to extend datasets by providing plausible artificially generated climate data (statistical robustness and rare events, including effects of a changing climate).

NOTES

1. Only two dimensions are shown, the Y-axis (not shown) is perpendicular to the page.
2. Many environmental properties, while observable in principle, cannot be observed well in practice. A good example is the deep ocean interior which is very sparsely sampled observationally.

3. This response to boundary condition errors also acts to *limit the predictability* for the evolution any single trajectory run from a specific set of initial conditions.
4. It should be noted that, like GCM ensembles, NWP ensembles may include sampling of model and parameter uncertainty. Indeed, recent developments have seen NWP and GCM models begin to converge in many respects, as NWP models include more Earth system components (e.g., coupling the atmosphere to ocean models) and the grid-resolution of GCMs increases.

References

Deser, C., Phillips, A., Bourdette, V., & Teng, H. (2012). Uncertainty in climate change projections: the role of internal variability. *Climate Dynamics, 38,* 527–546.

Lorenz, E. N. (1963). Deterministic non-periodic flow. *Journal of the Atmospheric Sciences, 20,* 130–141.

Lorenz, E. N. (1975). Climate predictability. In B. Bolin, et al. (Eds.), *The physical basis of climate and climate modelling* (Vol. 16, pp. 132–136). GARP Publication Series. Geneva: WMO.

Meehl, G. A., Stocker, T. F., Collins, W. D., Friedlingstein, P., Gaye, A. T., Gregory, J. M., et al. (2007). Global climate projections. In Solomon, S., D. Qin, M. Manning, Z. Chen, M. Marquis, K. B. Averyt, M. Tignor, & H. L. Miller (Eds.), *Climate change 2007: The physical science basis. Contribution of working group I to the fourth assessment report of the intergovernmental panel on climate change.* Cambridge; New York: Cambridge University Press.

Palmer, T. N. (1999). A nonlinear dynamical perspective on climate prediction. *Journal of Climate, 12,* 575–591.

Scaife, A. A., Arribas, A., Blockley, E., Brookshaw, A., Clark, R. T., Dunstone, N., et al. (2014). Skillful long-range prediction of European and North American winters. *Geophysical Research Letters, 41,* 2514–2519.

Schneider, T., & Griffies, S. M. (1999). A conceptual framework for predictability studies. *Journal of Climate, 12,* 3133–3155.

Smagorinsky, J., Manabe, S., & Holloway, J. L. (1965). Numerical results from a nine-level general circulation model of the atmosphere. *Monthly Weather Review, 93,* 727–768.

Stainforth, D. A., Aina, T., Christensen, C., Collins, M., Faull, N., Frame, D. J., et al. (2005). Uncertainty in predictions of the climate response to rising levels of greenhouse gases. *Nature, 433,* 403–406.

Taylor, K. E., Stouffer, R. J., & Meehl, G. A. (2012). An overview of CMIP5 and the experiment design. *Bulletin of the American Meteorological Society, 93,* 485–498.

Open Access This chapter is distributed under the terms of the Creative Commons Attribution 4.0 International License (http://creativecommons.org/licenses/by/4.0/), which permits use, duplication, adaptation, distribution and reproduction in any medium or format, as long as you give appropriate credit to the original author(s) and the source, a link is provided to the Creative Commons license and any changes made are indicated.

The images or other third party material in this chapter are included in the work's Creative Commons license, unless indicated otherwise in the credit line; if such material is not included in the work's Creative Commons license and the respective action is not permitted by statutory regulation, users will need to obtain permission from the license holder to duplicate, adapt or reproduce the material.

CHAPTER 7

Short-Range Forecasting for Energy

Sue Ellen Haupt

Abstract Short-range forecasts for periods on the order of hours to days and up to two weeks ahead are necessary to smoothly run transmission and distribution systems, plan maintenance, protect infrastructure and allocate units. In particular, forecasting the renewable energy resources on a day-to-day basis enables integration of increasing capacities of these variable resources. This chapter describes the basics of this short-range forecasting, beginning with the observation-based "nowcasting" of the first 15 minutes and ranging up to two weeks using numerical weather prediction. We discuss how blending multiple forecasts can increase accuracy and how probabilistic forecasts are being used to quantify the forecast uncertainty.

Keywords Wind power forecasts • Solar power forecasts • Renewable energy • Nowcasting • Numerical weather prediction • DICast • Forecast blending • Analog ensemble • Probabilistic forecasts

S.E. Haupt (✉)
National Center for Atmospheric Research (NCAR),
Boulder, CO, USA

© The Author(s) 2018   97
A. Troccoli (ed.), *Weather & Climate Services for the Energy Industry*,
https://doi.org/10.1007/978-3-319-68418-5_7

THE NEED FOR SHORT-RANGE FORECASTS

Utilities and Independent Transmission Operators (ITOs) depend on accurate forecasts for the next hours to several weeks. They often need to plan operation and maintenance outages weeks to months ahead. Weather events may make it difficult to maintain power lines, wind turbines and other infrastructure. If they can plan around events such as lightning storms and high wind events, the safety of the workers and efficiency of the maintenance can be greatly improved.

Weather also impacts day ahead planning of how to commit units. Although weather impacts all types of power production, it actually also drives the renewable units. It is important to be able to forecast the wind, solar and hydro power available the next day, or often, over the next several days too. The marginal cost to run these renewable resources is quite low and it is economically advantageous to allocate as much power from those units as possible. But overallocation of those units when the wind, irradiance or water power is not available could lead to using much more expensive reserve units in real time. Thus it is critical to produce high quality wind and solar power forecasts.

For timescales less than a day, it is important to know the most recent update to the forecast in order to balance the load in close to real time. Sharp up or down ramps in renewable energy must be balanced with the reserve units if the grid operators were not expecting them.

OVERVIEW OF SCALES

To forecast across these scales from minutes to a few weeks, meteorologists must combine forecast methods that are most appropriate to each scale. Figure 7.1 illustrates the big picture of the types of systems that may be useful for such forecasts. Figure 7.1 focuses on the short time range where the stakeholders require optimal accuracy. That figure demonstrates that observation-based nowcasting, which refers to the first few hours of the forecast, provides a much more accurate forecast in the short range, but its skill drops off rapidly. Numerical weather prediction (NWP) becomes more important at about three hours and provides value out to about two weeks. Because NWP run at high resolution over a sizable domain requires on the order of hours to run on supercomputers and often requires spin-up time, it is not available for real-time use in the shortest ranges. Modern methods of forecasting renewable energy output employing postprocessing methods to blend disparate models or ensem-

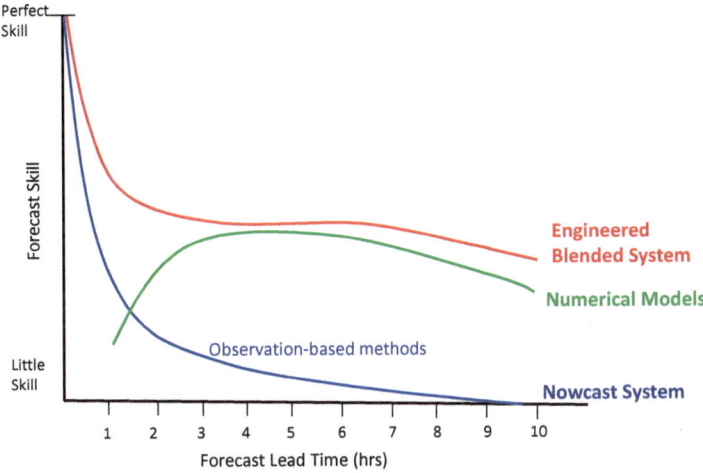

Fig. 7.1 Blending of NWP models with observation-based nowcasting enables optimization of the short-range forecast

bles are shown to greatly improve the forecast skill (Giebel and Kariniotakis 2007; Monteiro et al. 2009; Mahoney et al. 2012; Ahlstrom et al. 2013; Orwig et al. 2014; Tuohy et al. 2015). Figure 7.1 indicates that this blending can provide value beyond that of the input models.

Nowcasting

The shortest time frames of the first several hours benefit from forecasting based on data sensors that determine the current situation of the variable of interest. For instance, having *in situ* measurements of wind speed or solar radiation in the field allow us to train statistical learning or artificial intelligence models to recognize current conditions and likely changes. Over the first 15–45 minutes, it is often difficult to beat a persistence forecast. If we know the wind now, the best forecast in the first few minutes is that there is no change. For solar radiation, we often use a "smart persistence" where we expect the cloud cover to stay the same, but recognize that the solar angle will change. For solar energy, sky imaging traces cloud conditions in real time, which allows us to use motion vectors derived from image processing succeeding images, or if there is a co-located wind profiler, one can project the cloud motion for the next few steps (Kleissl 2010; Chu et al. 2013; Nguyen and Kleissl 2014; Peng et al. 2015). Cloud imager-based forecasts are effective for about 15–30 minutes.

Statistical learning methods have been shown to be effective from about 30 minutes to around three hours. Such methods may consist of using techniques such as artificial neural networks (ANN—Mellit 2008; Wang et al. 2012), autoregressive models (Hassanzadeh et al. 2010; Yang et al. 2012), Markov process models (Morf 2014) or support vector machines (Sharma et al. 2011; Bouzerdoum et al. 2013) to recognize patterns in the changes of wind speed or solar radiation. Blended techniques, such as that of Pedro and Coimbra (2012), that use a genetic algorithm to optimize an ANN have also been effective. In addition, new methods blend weather observations with irradiance observations and use clustering techniques to identify regimes and then train ANN for the individual regimes (Kazor and Hering 2015), which was shown to improve upon non-regime-dependent ANN and upon smart persistence (McCandless et al. 2016a). Recent work has added satellite data to this type of forecasting (McCandless et al. 2016b). Some methods also predict the variability of the resource (McCandless et al. 2015).

In the time range from about an hour out to six hours, cloud motion vectors derived from satellite data are often used to forecast solar irradiance (Miller et al. 2013). When a satellite observes the cloud cover and the forecast system then advects it using motion vectors derived from successive images or uses observed or modeled data to advect those clouds, the derived liquid water path can be used to provide good estimates of irradiance attenuation. Satellite-based methods depend on the satellite data being received and processed before being used in the motion vector models, which take on the order of 30 minutes after observation. They also do not account for cloud development or dissipation, so the forecasts of individual clouds are only accurate for a limited period of time, beginning to degrade after the first couple hours.

Wind energy forecasting at these scales of the first few hours can be improved by other methods of remote sensing. Mahoney et al. (2012) describe the Variational Doppler Radar Analysis System (VDRAS) that assimilates radar data into a cloud-resolving model to better predict winds. Because that model does not include the full physics, it can be updated at frequencies as high as every 15 minutes. That work showed that in case studies, the winds could be well predicted for the first two hours and could identify weather ramps as they approach. This method, however, relies on having radars sited in close proximity to the wind farms.

Numerical Weather Prediction

Beyond about three to six hours, the workhorse of forecasting is NWP. NWP consists of the integration of the nonlinear partial differential equations governing atmospheric flow and includes appropriate models for the physics of clouds, radiation, turbulence, land surface conditions and more (Warner 2011). As computer power has advanced, so has our capability to provide higher resolution simulations in closer to real time. The national centers now run very short-range simulations at about three kilometers horizontal resolution over limited regions (such as over the USA) as often as hourly. Global models have necessarily coarser resolution and run less frequently. As of mid-2017, the US Global Forecast System is run every six hours at 13 kilometers resolution with hourly output for the first five days and at 70 kilometers out to 16 days. The European Center for Medium Forecast is run twice a day at nine kilometers resolution for 16 days with output at three hours temporal resolution. The national centers are continually updating their model resolutions, lengths of simulations and frequency of the runs as computer power is upgraded. In addition, they are including ensemble runs, which provide probabilistic information as well as improving upon the deterministic forecast. As mentioned in the introduction and elaborated later, the sensitivity to initial conditions is what partly limits predictability, which necessitates running ensembles of models.

It is important to assimilate observations of weather data from the global networks to provide the best possible initial condition to the runs. To improve forecasts at specific points, such as at a wind farm, it is advantageous to also assimilate specialized data (such as wind speed measurements) at that farm. Mahoney et al. (2012) and Wilczak et al. (2015) provide evidence that assimilating local wind farm data can improve the NWP forecasts. In a case study, Cheng et al. (2017) show that real-time four-dimensional data assimilation can reduce the mean absolute error in the forecast by 30–40% in the first three hours. Kosovic et al. (2017) indicate that such local data assimilation can significantly improve forecasts over the national models for the first 15 hours. Versions of the models are now being built to specifically improve upon predicting variables of particular importance to energy, such as wind (Wilczak et al. 2015), solar (Jimenez et al. 2016) and hydro (Gochis et al. 2014), although the use of national hydrological models is still in its infancy.

Blending the Forecasts and Predicting Power

Modern forecasting includes postprocessing the NWP output and forecast blending to improve upon the results. At a basic level, a multivariate statistical regression model known as model output statistics (MOS) is applied to remove biases (Glahn and Lowry 1972). More complex methods (such as ANN, autoregressive models and others) are also used to provide nonlinear corrections to models (Myers et al. 2011; Myers and Linden 2011; Giebel and Kariniotakis 2007; Pelland et al. 2013). Ensemble MOS (Wilks and Hamill 2007) not only corrects the individual models but also optimizes weights for blending forecasts from multiple models. It is typical to blend multiple models together using some statistical learning technique to produce a better forecast than any single model could produce consistently, often with a 10–15% improvement over the best model forecast (Mahoney et al. 2012; Myers et al. 2011). Figure 7.2 displays an example of wind speed forecasts using National Center for Atmospheric Research (NCAR's) Dynamic Integrated foreCast (DICast®) system to blend multiple models, with the blended forecast showing a lower average root mean square error than any input model.

Probabilistic Forecasts and the Analog Ensemble

Utilities and ITOs are also requesting probabilistic information to plan reserves, alleviating transmission bottlenecks, and better planning for renewable operations. To this end, the meteorology sector has been providing probabilistic forecasts. All major national forecasting centers now run ensemble forecasts. The rate of spread of the members of the ensemble quantifies the uncertainty. Various postprocessing methods are used to remove bias, sharpen the probability density function and calibrate the spread, or reliability, of the forecast.

Another interesting postprocessing method is the analog ensemble (AnEn) technique. In this case, however, rather than running multiple models, the historical output from a single, often high resolution, model is used to generate the ensemble. A search of the historical forecast records is made to identify the forecasts that are most similar to the current forecast. We then compare the forecasts to the corresponding observations. Those observations then become the analog ensemble. This method is effective at both improving on the deterministic forecast and using the multiple analogs to form an ensemble that can be used to quantify the

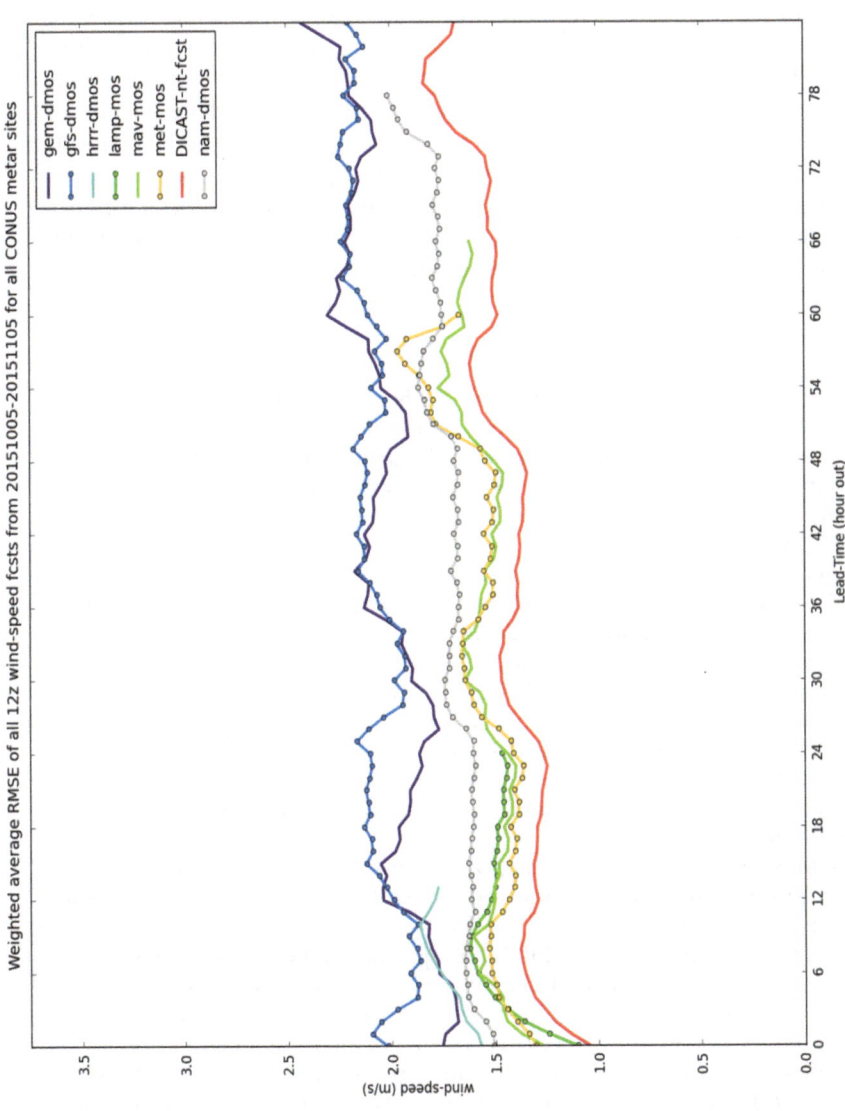

Fig. 7.2 Mean root mean square error for wind speed forecasts at METAR sites over the contiguous USA from multiple models and the DICast forecast (red) for the month from 5 October to 5 November, 2015

uncertainty of the forecast (Delle Monache et al. 2013; Haupt and Delle Monache 2014; Alessandrini et al. 2015).

Accuracy of forecasts has been improving steadily with some areas now seeing single digit errors in terms of percentage of capacity at a wind farm (Orwig et al. 2014; Haupt and Mahoney 2015). These improvements have stemmed from including observations in the immediate vicinity of the resource, both in the nowcasting and assimilated into the NWP models, as well as better methods of blending multiple models for the appropriate timescales. Solar power predictions have not been a focus for very long, but rapid improvement is also happening here (Lorenz et al. 2009; 2014; Tuohy et al. 2015; Haupt et al. 2017).

REFERENCES

Ahlstrom, M., Bartlett, D., Collier, C., Duchesne, J., Edelson, D., Gesino, A., et al. (2013). Knowledge is power: Efficiently integrating wind energy and wind forecasts. *IEEE Power and Energy Magazine, 11*(6), 45–52.

Alessandrini, S., Delle Monache, L., Sperati, S., & Cervone, G. (2015). An analog ensemble for short-term probabilistic solar power forecast. *Applied Energy, 157,* 95–110.

Bouzerdoum, M., Mellit, A., & Pavan, A. (2013). A hybrid model (SARIMA-SVM) for short-term power forecasting of a small-scale grid-connected photovoltaic plant. *Solar Energy, 98,* 226–235.

Cheng, W. Y. Y., Liu, Y., Bourgeois, A., Wu, Y., & Haupt, S. E. (2017). Short-term wind forecast of a data assimilation/weather forecasting system with wind turbine anemometer measurement assimilation. *Renewable Energy, 107,* 340–351. https://doi.org/10.1016/j.renene.2017.02.014.

Chu, Y., Pedro, H., & Coimbra, C. (2013). Hybrid intra-hour DNI forecasts with sky image processing enhanced by stochastic learning. *Solar Energy, 98,* 592–603.

Delle Monache, L., Eckel, F. A., Rife, D. L., Nagarajan, B., & Searight, K. (2013). Probabilistic weather prediction with an Analog Ensemble. *Monthly Weather Review, 141,* 3498–3516.

Giebel, G., & Kariniotakis, G. (2007). Best practice in short-term forecasting – A users guide. European wind energy conference and exhibition. Milan (IT), 7–10 May 2007. Retrieved from http://powwow.risoe.dk/publ.htm

Glahn, H. R., & Lowry, D. A. (1972). The use of model output statistics (MOS) in objective weather forecasting. *Journal of Applied Meteorology, 11,* 1203–1211.

Gochis, D. J., Yu, W., & Yates, D. N. (2014). The WRF-Hydro model technical description and user's guide, version 2.0. NCAR Technical Document. 120 pages. Retrieved from http://www.ral.ucar.edu/projects/wrf_hydro/

Hassanzadeh, M., Etezadi-Amoli, M., & Fadali, M. S. (2010). Practical approach for sub-hourly and hourly prediction of PV power output. *North American Power Symposium (NAPS)*, 1–5, September 26–28.

Haupt, S. E., & Delle Monache, L. (2014). Understanding ensemble prediction: How probabilistic wind power prediction can help in optimizing operations. *WindTech, 10*(6), 27–29.

Haupt, S. E., & Mahoney, W. P. (2015). Wind power forecasting. *IEEE Spectrum, 2015,* 46–52.

Haupt, S. E., Kosović, B., Jensen, T., Lazo, J. K., Lee, J. A., Jiménez, P. A., et al. (2017). Building the Sun4Cast system: Improvements in solar power forecasting. *Bulletin of the American Meteorological Society.* https://doi.org/10.1175/BAMS-D-16-0221.1.

Jimenez, P. A., Hacker, J. P., Dudhia, J., Haupt, S. E., Ruiz-Arias, J. A., Gueymard, C. A., et al. (2016). WRF-solar: An augmented NWP model for solar power prediction. *Bulletin of the American Meteorological Society,* 1249–1264. https://doi.org/10.1175/BAMS-D-14-00279.1.

Kazor, K., & Hering, A. S. (2015). Assessing the performance of model-based clustering methods in multivariate time series with application to identifying regional wind regimes. *Journal of Agricultural, Biological, and Environmental Statistics, 20,* 192–217.

Kleissl, J. (2010). *Current state of the art in solar forecasting.* Final Report California Renewable Energy Forecasting, Resource Data and Mapping Appendix A. California Renewable Energy Collaborative.

Kosovic, B., Haupt, S. E., Adriaansen, D., Alessandrini, S., Jensen, T., Wiener, G., et al. (2017). Scientific advances in wind power forecasting. Submitted to *Wind Energy.*

Lorenz, E., Hurka, J., Heinemann, D., & Beyer, H. (2009). Irradiance forecasting for the power prediction of grid-connected photovoltaic systems. *IEEE Journal of Selected Topics in Applied Earth Observations and Remote Sensing, 2*(1), 2–10.

Lorenz, E., Kihnert, I., & Heinemann, D. (2014). Overview of irradiance and photovoltaic power prediction. In A. Troccoli, L. Dubus, & S. E. Haupt (Eds.), *Weather matters for energy* (pp. 420–454). New York: Springer.

Mahoney, W. P., Parks, K., Wiener, G., Liu, Y., Myers, B., Sun, J., et al. (2012). A wind power forecasting system to optimize grid integration. Special issue of *IEEE Transactions on Sustainable Energy* on Applications of Wind Energy to Power Systems, *3* (4), 670–682.

McCandless, T. C., Haupt, S. E., & Young, G. S. (2015). A model tree approach to forecasting solar irradiance variability. *Solar Energy, 120,* 514–524. https://doi.org/10.1016/j.solener.2015.07.0200038-092X.

McCandless, T. C., Haupt, S. E., & Young, G. S. (2016a). A regime-dependent artificial neural network technique for short-range solar irradiance forecasting. *Applied Energy, 89,* 351–359.

McCandless, T. C., Young, G. S., Haupt, S. E., & Hinkelman, L. M. (2016b). Regime-Dependent short-range solar irradiance forecasting. *Journal of Applied Meteorology and Climatology, 55*, 1599–1613.

Mellit, A. (2008). Artificial intelligence technique for modeling and forecasting of solar radiation data: A review. *Int. Journal Artificial Intelligence and Soft Computing, 1*(1), 52–76.

Miller, S. D., Heidinger, A. K., & Sengupta, M. (2013). Chapter 3: Physically based satellite methods. In J. Kleissl (Ed.), *Solar energy forecasting* (504 pp.). Oxford: Elsevier. ISBN:9780123971777.

Monteiro, C., Bessa, R., Miranda, V., Botterud, A., Wang, J., & Conzelmann, G.. (2009). Wind power forecasting: State-of-the-Art 2009. Argonne National Laboratory, Argonne, IL, ANL/DIS-10-1, November 2009.

Morf, H. (2014). Sunshine and cloud cover prediction based on Markov processes. *Solar Energy, 110*, 615–626.

Myers, W., & Linden, S. (2011). *A turbine hub height wind speed consensus forecasting system.* AMS Ninth Conference on Artificial Intelligence and its Applications to the Environmental Sciences, Seattle, WA, January 23–27, 2011.

Myers, W., Wiener, G., Linden, S., & Haupt, S. E. (2011). *A consensus forecasting approach for improved turbine hub height wind speed predictions.* Proceedings of WindPower 2011.

Nguyen, D., & Kleissl, J. (2014). Stereographic methods for cloud base height determination using two sky imagers. *Solar Energy, 107*, 495–509.

Orwig, K. D., Ahlstrom, M., Banunarayanan, V., Sharp, J., Wilczak, J. M., Freedman, J., et al. (2014). Recent trends in variable generation forecasting and its value to the power system. *IEEE Transactions on Renewable Energy, 6*(3), 924–933.

Pedro, H. T., & Coimbra, C. F. M. (2012). Assessment of forecasting techniques for solar power production with no exogenous inputs. *Solar Energy, 86*(7), 2017–2028.

Pelland, S., Remund, J., Kleissl, J., Oozeki, T., & De Brabandere, K.. (2013). Photovoltaic and solar forecasting: State of the art. IEA PVPS Task 14, Subtask 3.1 Report IEA-PVPS T14-01: 2013.

Peng, Z., Yu, D., Huang, D., Heiser, J., Yoo, S., & Kalb, P. (2015). 3D cloud detecting and tracking system for solar forecast using multiple sky imagers. *Solar Energy, 118*, 496–519.

Sharma, N., Sharma, P., Irwin, D., & Shenoy, P. (2011). *Predicting solar generation from weather forecasts using machine learning.* Proceedings of the 2nd IEEE International Conference on Smart Grid Communications, Brussels, 17–20 October, pp. 32–37.

Tuohy, A., Zack, J., Haupt, S., Sharp, J., Ahlstrom, M., Dise, S., et al. (2015). Solar forecasting – Method, challenges, and performance. *IEEE Power and Energy Magazine, 13*(6), 50–59. https://doi.org/10.1109/MPE.2015.2461351.

Wang, F., Mi, Z., Su, S., & Zhao, H. (2012). Short-term solar irradiance forecasting model based on artificial neural network using statistical feature parameters. *Energies, 5*, 1355–1370.

Warner, T. T. (2011). *Numerical weather and climate prediction*. Cambridge: Cambridge University Press. 526 pp.

Wilczak, J., et al. (2015). The Wind Forecast Improvement Project (WFIP): A Public-Private Partnership addressing wind energy forecast needs. *Bulletin of the American Meteorological Society, 96*, 1699–1718.

Wilks, D. S., & Hamill, T. M. (2007). Comparison of ensemble-MOS methods using GFS reforecasts. *Monthly Weather Review, 135*, 2379–2390.

Yang, D., Jirutitijaroen, P., & Walsh, W. M. (2012). Hourly solar irradiance time series forecasting using cloud cover index. *Solar Energy, 86*, 3531–3543.

Open Access This chapter is distributed under the terms of the Creative Commons Attribution 4.0 International License (http://creativecommons.org/licenses/by/4.0/), which permits use, duplication, adaptation, distribution and reproduction in any medium or format, as long as you give appropriate credit to the original author(s) and the source, a link is provided to the Creative Commons license and any changes made are indicated.

The images or other third party material in this chapter are included in the work's Creative Commons license, unless indicated otherwise in the credit line; if such material is not included in the work's Creative Commons license and the respective action is not permitted by statutory regulation, users will need to obtain permission from the license holder to duplicate, adapt or reproduce the material.

CHAPTER 8

Medium- and Extended-Range Ensemble Weather Forecasting

David Richardson

Abstract The chapter provides an overview of ensemble weather forecasting for the medium- and extended-range (days to weeks ahead). It reviews the methods used to account for uncertainties in the initial conditions and in the forecast models themselves. The chapter explores the challenges of making useful forecasts for the sub-seasonal timescale, beyond the typical limit for skilful day-to-day forecasts, and considers some of the sources of predictability such as the Madden-Julian oscillation (MJO) that make this possible. It then introduces some of the ensemble-based forecast products and concludes with a case study for a European heat wave that demonstrates how ensemble weather forecasts can be used to guide decision making for weather-dependent activities.

Keywords Ensemble • Weather forecast • Uncertainty • Predictability • Medium-range • Extended-range • Sub-seasonal • Madden-Julian oscillation

D. Richardson (✉)
European Centre for Medium-Range Weather Forecasts (ECMWF),
Reading, UK

© The Author(s) 2018
A. Troccoli (ed.), *Weather & Climate Services for the Energy Industry*,
https://doi.org/10.1007/978-3-319-68418-5_8

PREAMBLE

Operational weather forecasts for the medium- and extended-range (days to weeks ahead) are generally based on the output from global Numerical Weather Prediction (NWP) or General Circulation Model (GCM) ensemble forecasts.

INITIAL CONDITION UNCERTAINTIES

The main aim of the ensemble approach is to account for uncertainty in the initial atmospheric conditions. Typically this is done by adding small perturbations to a single "best-estimate" analysis of the current state of the atmosphere. The analysis is generated through assimilation of observations into the NWP model. The size of the perturbations is constrained to be consistent with the known analysis errors. A number of different methodologies are used to generate perturbations that are physically realistic and that will grow to represent the range of possible future states consistent with the initial uncertainty.

The error-breeding method (Toth and Kalnay 1993, 1997) uses a cycling approach where the differences between short-range forecasts are re-scaled to form the initial perturbations for the next forecast. A number of generalisations of the original breeding method have been developed, designed to improve the representation of the analysis uncertainty at each initial time. These include the Ensemble Transform Kalman Filter (ETKF, Bishop et al. 2001), used at the Met Office (Bowler et al. 2008, 2009), and the Ensemble Transform with Rescaling (ETR, Wei et al. 2008) used at the United States National Centers for Environmental Prediction (NCEP).

The Singular Vector (SV) method (Buizza and Palmer 1995) computes new perturbations at each analysis time. The method identifies the fastest growing perturbations over a given time period (e.g. 48 hours). A linear combination of these SVs, scaled to have amplitudes consistent with the analysis error, is added to the best-estimate analysis to make the starting conditions for each ensemble member. The SV method is used operationally at the European Centre for Medium-Range Weather Forecasts (ECMWF) (Leutbecher and Palmer 2008) and the Japan Meteorological Agency (JMA) (Yamaguchi and Majumdar 2010).

Other methods more directly address the observation uncertainty by perturbing the observed values themselves. At ECMWF, perturbations are also provided from an Ensemble of Data Assimilations (EDA) (Buizza et al. 2008). Each EDA member is an independent data assimilation, using the same set of observations, but introducing perturbations to these observations consistent with the known observation errors. The Meteorological Service of Canada (MSC) uses perturbed observations and an ensemble approach, the ensemble Kalman filter (EnKF, Houtekamer and Mitchell 2005; Houtekamer et al. 2009, 2014), to provide an ensemble of initial conditions. It should be noted that for both the EDA and EnKF, it is necessary to take account of model uncertainties (see below) as well as the observation uncertainties to generate appropriate initial perturbations.

MODEL UNCERTAINTIES

Global NWP ensemble forecasts typically run with a grid spacing of a few tens of kilometres. Many important physical processes (that affect, e.g., clouds and precipitation) work on much smaller spatial scales than can be resolved directly. These processes are represented in the NWP models by "parametrization schemes" that describe the aggregate effect of the smaller-scale unresolved processes on the larger resolved scales.

The finite resolution of the NWP model and the approximations made in the parametrisation schemes are sources of model uncertainty. Most global NWP ensembles also now include a representation of these uncertainties in the model formulation. A range of methods has been developed and sometimes a combination of methods is used in a single ensemble system to account for different aspects of model uncertainty.

One approach is to use a number of different parametrisation schemes within the ensemble. For example, there are various ways to parametrise convective processes and an ensemble can be generated by running some members using one convection parametrisation scheme, while other members use a different convection scheme (Charron et al. 2010). An alternative is to use a single parametrisation, but to perturb some of the key parameters in the scheme (Bowler et al. 2008). Other schemes represent the uncertainty from the sub-grid scale by stochastically perturbing the tendencies from the parametrisation schemes, as, for example, in the Stochastically Perturbed Parametrisation Tendency scheme (SPPT, Leutbecher et al. 2017; Buizza et al. 1999). Backscatter schemes are designed to simulate the transfer of energy from the unresolved sub-grid scales to the larger scales that are resolved by the model (Shutts 2005; Berner et al. 2009).

OPERATIONAL GLOBAL MEDIUM-RANGE ENSEMBLES

A number of meteorological centres produce operational medium-range ensemble forecasts. Initial condition and model uncertainties are represented using many of the above methods, with different centres adopting different approaches. The TIGGE (The International Grand Global Ensemble) project provides access to regular global ensemble predictions from ten of the leading global NWP centres to support research and has facilitated a comprehensive evaluation of the global ensembles produced by different NWP centres (Swinbank et al. 2016; Bougeault et al. 2010; and references therein).

Comparison of the forecasts from the TIGGE centres confirms that while the different centres each have their strengths and weaknesses, the different perturbation methodologies all have merit. It is more important that an NWP system produces an ensemble that accounts for both initial condition and model uncertainties than the precise methodology used to produce the perturbations. However, it is also important to carry out proper and comprehensive evaluation to ensure that the perturbations are consistent with the uncertainties of the system. Figure 8.1 shows an example of the skill of operational ensemble forecasts from five global centres in predicting the large-scale weather patterns over the extra-tropical northern hemisphere up to two weeks ahead (the temperature at 850 hPa is a good indicator of whether a location is under the influence of a warm or cool circulation pattern). Skill is measured using the Continuous Ranked Probability Skill Score (CRPSS), a standard measure for assessing the usefulness of probabilistic forecasts, which can also be interpreted as an indication of the potential economic value of the forecast systems (Palmer and Richardson 2014). CRPSS ranges from a maximum value of 1 (perfect knowledge of what the weather will be) to zero (only the climatological information is known). This evaluation shows that all the forecasting systems have positive skill in forecasting day-to-day changes in the weather for up to two weeks ahead.

EXTENDED-RANGE ENSEMBLES

The medium-range ensembles described above typically produce forecasts for one to two weeks ahead. This is usually considered the limit for day-to-day predictability, as the influence of the atmospheric initial conditions is much reduced at longer range.

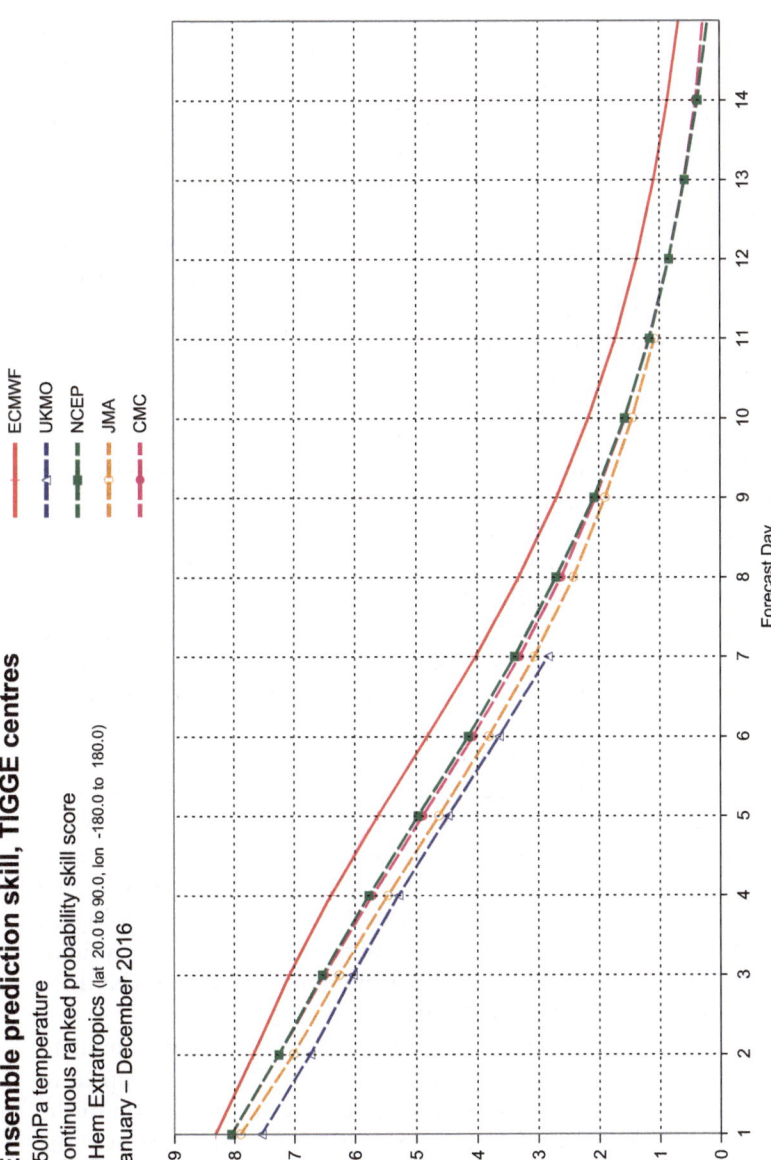

Fig. 8.1 Skill of ensemble forecasts for temperature at 850 hPa in the northern hemisphere extra-tropics for 2016. The verification is performed against each centre's own analysis, with the forecast and analysis data taken from the TIGGE archive. CMC = Canadian Meteorological Centre, JMA = Japan Meteorological Agency, UKMO = United Kingdom Met Office; NCEP = The National Centres for Environmental Prediction (USA)

However, there are important sources of predictability that do give the potential to make useful forecasts for the sub-seasonal scale (one or two months ahead). One important phenomenon is the Madden-Julian oscillation (MJO), a feature of the tropical atmosphere that also impacts on the weather in the extra-tropics (Lin et al. 2009; Cassou 2008). The MJO evolves over a period of 40–60 days, and so is a potential source of forecast skill for several weeks ahead. Initial conditions in the stratosphere can affect the circulation in the troposphere over the following month, providing another source for sub-seasonal predictability (Baldwin and Dunkerton 2001). Other potential sources of predictability at this timescale include the land surface (Koster et al. 2010) and snow cover (Jeong et al. 2013) conditions at the start of the forecast.

There have been significant improvements in sub-seasonal forecasts in recent years, with large improvements in skill for predicting the MJO (Fig. 8.2), as well as its influence on other regions, including Europe (Vitart 2014). There is now a growing interest in developing applications to exploit these forecasts as well as to improve the forecasts themselves.

Following the success of TIGGE, a new sub-seasonal to seasonal prediction project (S2S) has been initiated by the World Weather Research Programme (WWRP) and World Climate Research Programme (WCRP). The main goal of this five-year project is to improve forecast skill and understanding of the sub-seasonal to seasonal timescale and to promote its uptake by operational centres and its exploitation by the applications community (Vitart et al. 2012).

The S2S database includes near real-time ensemble forecasts for up to 60 days ahead, from 11 forecasting centres: Australia's Bureau of Meteorology (BOM); the China Meteorological Administration (CMA); ECMWF; Environment and Climate Change Canada (ECCC); Italy's Institute of Atmospheric Sciences and Climate (CNR-ISAC); the Hydrometeorological Centre of Russia (HMCR); the Japan Meteorological Agency (JMA); the Korea Meteorological Administration (KMA); Météo-France; the US National Centers for Environmental Prediction (NCEP); and the UK Met Office (Vitart et al. 2017).

These models are generally different from the NWP models used to produce medium-range forecasts at the same centres. Most are coupled to an ocean model, as it is important to take account of the evolution of the sea-surface temperature and its interaction with the atmosphere over the longer time periods of these forecasts. For the same reason, some systems also include an active sea ice model. Some centres, such as ECMWF, that

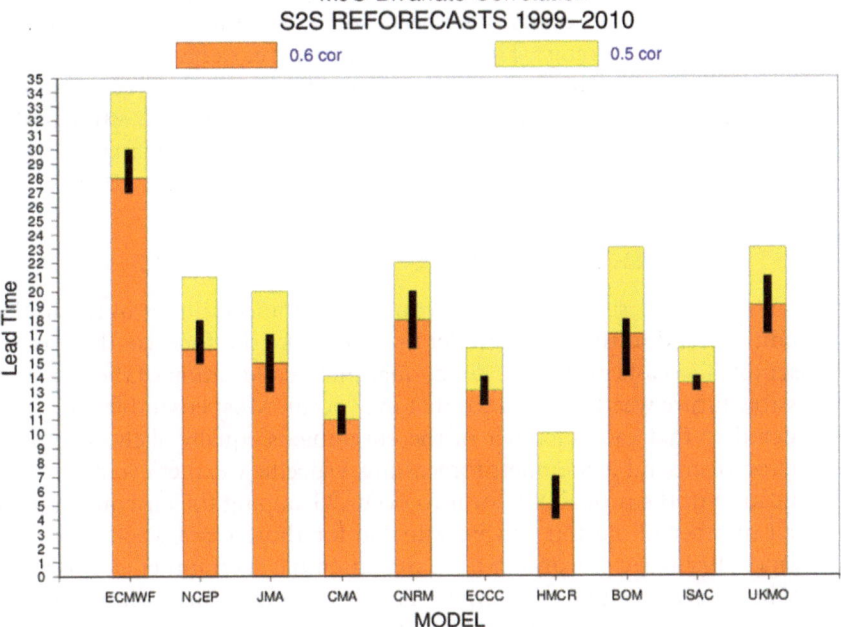

Fig. 8.2 Forecast lead-time (in days) when a correlation-based measure of accuracy of the prediction of the Madden-Julian Oscillation (MJO) reaches 0.6 correlation (orange bars) and 0.5 correlation (yellow bars) (1.0 would indicate a perfect forecast). The black lines indicate the 95% confidence interval of the time when the 0.6 correlation is reached. Results are based on the re-forecast from 1999 to 2010 from all the models, verified against ERA-Interim analyses. Correlations of 0.5 and 0.6 are often used as indication of useful forecast skill (Vitart 2014)

already include an ocean model in the medium-range ensemble, use the same ensemble system for both the medium-range and sub-seasonal forecasts, hence providing a seamless set of predictions covering the timescales from days to months ahead. While the skill in predicting the day-to-day weather may be limited to around two weeks in general (as seen in Fig. 8.1), many of the S2S models demonstrate substantial skill out to three or even four weeks ahead for the MJO (Fig. 8.2); this can lead to enhanced predictability, for example, over Europe, in certain situations.

Another important aspect of the sub-seasonal forecasts is the need to account for model errors. Systematic model errors (biases) can accumulate during the forecast, and while they are often small enough to be neglected for medium-range forecasts, they become too large to be ignored at the longer sub-seasonal range. Hence, an additional set of historical ensemble integrations is generated by making forecasts from start dates covering the last 15–20 years. These re-forecasts (or hindcasts) are used to estimate the model climate. This can then be used to remove the model biases from the real-time forecasts in a statistical post-processing step (Vitart 2014).

ENSEMBLE WEATHER FORECAST PRODUCTS

Ensembles are designed to take account of the uncertainties in the initial conditions and in the NWP model used to make the forecast. The set of ensemble forecasts provides a direct quantitative indication of the range of possible future weather scenarios that may occur. Most ensembles are constructed so that each member of the ensemble is equally likely. The proportion of ensemble members forecasting a specific weather event gives an indication of the probability for it to occur. Grouping the ensemble into a small number of clusters can be valuable for those cases when there are distinct alternative scenarios within the ensemble (Ferranti et al. 2015).

The forecast values of weather variables (temperature, wind, rainfall, etc.) are typically generated on a 10–50-kilometre spatial grid for medium- and extended-range ensembles. These values are not directly comparable to the measurements recorded at specific locations since they represent the average for the area covered by a grid-box. Statistical post-processing can substantially improve the forecasts by tailoring (down-scaling) the grid-box forecasts to smaller areas or individual sites, and also accounting for the finite ensemble size (Hemri et al. 2014). Re-forecasts are useful for calibrating medium-range forecasts as well as for the sub-seasonal range, and re-forecast datasets are increasingly becoming part of the medium-range forecast configurations. This can be especially important for severe weather forecasting by providing information about how the model performed for severe events in the re-forecast period.

The Extreme Forecast Index (EFI) was developed at ECMWF to highlight potential anomalous weather events, by comparing the real-time forecast to the re-forecast model climate distribution (Lalaurette 2003).

Beyond a few days ahead it is no longer possible to predict the day-to-day changes in the weather at specific locations. However, by considering the average conditions over a period of time, it is possible to give skilful forecasts for longer lead times (Buizza and Leutbecher 2015). Sub-seasonal forecasts typically predict average conditions for each week of the coming month. At this longer forecast range, the prediction of changes in large-scale weather patterns is important, for example, to give an early indication of the onset of heat waves or cold spells.

In summer 2015, a heat wave affected large parts of Europe—temperature records were broken in many places, including in Germany, France and Spain. Early indications of widespread warmer than normal conditions during the first week of July can be seen in the ECMWF forecast from 16 June (Fig. 8.3). The signal becomes noticeably stronger in the forecast

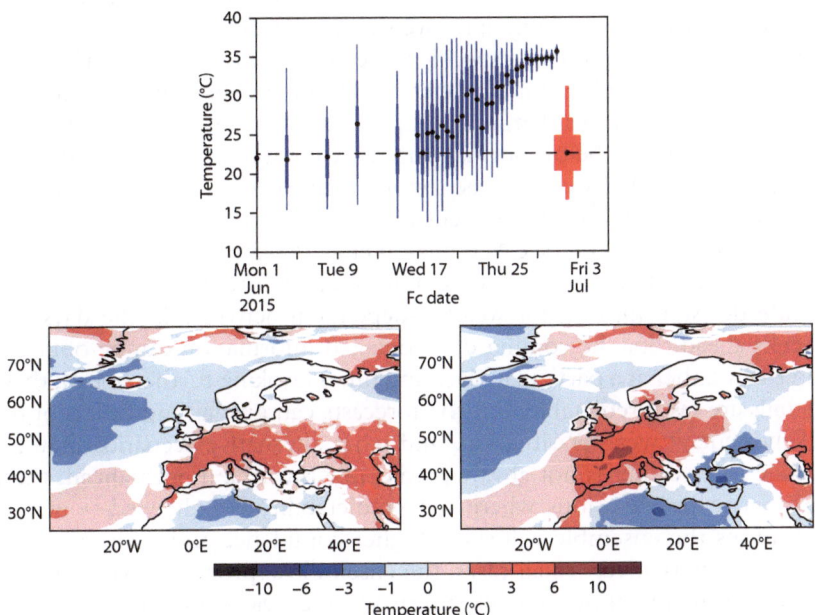

Fig. 8.3 ECMWF forecasts for the heat wave over Europe in July 2015. Lower panel shows the 2-metre temperature anomaly forecasts for the 7-day period 29 June to 5 July initialised on 18 June (left) and 22 June (right). Areas where the forecast distribution is significantly different from climatology are shaded. Upper panel shows the evolution of the ensemble forecasts for the temperature in Paris at 12 UTC on 1 July; the dates on the horizontal axis indicate the start time of each forecast. The box-and-whisker plots show the 1st, 10th, 25th, 75th, 90th and 99th percentile of the forecast, while black dot shows the median of the distribution. The temperature distribution of the model climate (generated from re-forecasts for late June and early July for the last 20 years) is shown in red (the dotted line highlights the climate median). Magnusson et al. 2015.

from 22 June. While these successive forecasts give a good indication of the general situation, it is also interesting to consider the forecast for a specific time and location. The top panel of Fig. 8.3 shows a sequence of forecasts for the temperature in Paris at 12 UTC on 1 July (this was the second warmest day on record for Paris, with temperatures reaching to almost 40°C later in the afternoon). Each forecast is represented as a box-and-whisker plot that summarises the information in the ensemble. The forecasts from early June indicate a range of possible outcomes, similar to the model climate distribution (shown in red), with no clear signal for temperature to be warmer or cooler than normal. This is not surprising: as noted in the previous paragraphs, we should not expect skill in predicting the precise temperature at a specific place and time out to a month ahead. By mid-June there is a noticeable shift in the forecasts: while not certain, the ensemble indicates that high temperatures are much more likely to be above normal than below normal, and there is a significant chance that the temperature in Paris could reach more than 30°C. The risk of extreme temperatures increases in consecutive forecasts, and by 26 June the outcome is almost certain (see Magnusson et al. 2015 for more details).

This example shows how ensemble weather forecasts can be used to guide decision making for weather-dependent activities in the days and weeks ahead. For the coming days, detailed information can be obtained on the weather at a particular place and time. Looking further ahead, these details are less predictable, but the forecasts can give an indication of the likely general weather situation, and what alternative scenarios may be. While the weather is not always predictable out to a month ahead, there are particular situations where the predictability is enhanced—on such occasions the ensemble will show higher confidence and a more limited range of alternatives. In all cases, the ensemble approach provides important information about possible alternative scenarios, and enables users to make appropriate decisions, taking account of the confidence and risks quantified by the ensemble.

References

Baldwin, M. P., & Dunkerton, T. J. (2001). Stratospheric harbingers of anomalous weather regimes. *Science, 294,* 581–584.

Berner, J., Shutts, G. J., Leutbecher, M., & Palmer, T. N. (2009). A spectral stochastic kinetic backscatter scheme and its impact on flow-dependent predictability in the ECMWF ensemble prediction system. *J. Atmos. Sci., 66,* 603–626.

Bishop, C. H., Etherton, B. J., & Majumdar, S. J. (2001). Adaptive sampling with the ensemble transform Kalman filter. Part I: theoretical aspects. *Monthly Weather Review, 129,* 420–436.

Bougeault, P., et al. (2010). The THORPEX Interactive Grand Global Ensemble (TIGGE). *Bulletin of the American Meteorological Society, 91,* 1059–1072.

Bowler, N. E., Arribas, A., Mylne, K. R., Robertson, K. B., & Beare, S. E. (2008). The MOGREPS short-range ensemble prediction system. *Quarterly Journal of the Royal Meteorological Society, 134,* 703–722. https://doi.org/10.1002/qj.234.

Bowler, N. E., Arribas, A., Beare, S. E., Mylne, K. R., & Shutts, G. J. (2009). The local ETKF and SKEB: Upgrades to the MOGREPS short-range ensemble prediction system. *Quarterly Journal of the Royal Meteorological Society, 135,* 767–776. https://doi.org/10.1002/qj.394.

Buizza, R., & Leutbecher, M. (2015). The Forecast skill horizon. *Quarterly Journal of the Royal Meteorological Society.* https://doi.org/10.1002/qj.2619.

Buizza, R., & Palmer, T. N. (1995). The singular vector structure of the atmospheric general circulation. *Journal of the Atmospheric Sciences, 52,* 1434–1456.

Buizza, R., Miller, M., & Palmer, T. N. (1999). Stochastic representation of model uncertainties in the ECMWF Ensemble Prediction System. *Quarterly Journal of the Royal Meteorological Society, 125,* 2887–2908.

Buizza, R., Leutbecher, M., & Isaksen, L. (2008). Potential use of an ensemble of analyses in the ECMWF Ensemble Prediction System. *Quarterly Journal of the Royal Meteorological Society, 134,* 2051–2066.

Cassou, C. (2008). Intraseasonal interaction between the Madden–Julian Oscillation and the North Atlantic Oscillation. *Nature, 455,* 523–527. https://doi.org/10.1038/nature07286.

Charron, M., Pellerin, G., Spacek, L., Houtekamer, P. L., Gagnon, N., Mitchell, H. L., et al. (2010). Toward random sampling of model error in the Canadian ensemble prediction system. *Monthly Weather Review, 138,* 1877–1901. https://doi.org/10.1175/2009MWR3187.1.

Ferranti, L., Corti, S., & Janousek, M. (2015). Flow-dependent verification of the ECMWF ensemble over the Euro-Atlantic sector. *Quarterly Journal of the Royal Meteorological Society, 141,* 916–924. https://doi.org/10.1002/qj.2411.

Hemri, S., Scheuerer, M., Pappenberger, F., Bogner, K., & Haiden, T. (2014). Trends in the predictive performance of raw ensemble weather forecasts. *Geophysical Research Letters, 41,* 9197–9205. https://doi.org/10.1002/2014GL062472.

Houtekamer, P. L., & Mitchell, H. L. (2005). Ensemble Kalman filtering. *Quarterly Journal of the Royal Meteorological Society, 131,* 3269–3289. https://doi.org/10.1256/qj.05.135.

Houtekamer, P. L., Mitchell, H. L., & Deng, X. (2009). Model error representation in an operational ensemble Kalman filter. *Monthly Weather Review, 137,* 2126–2143.

Houtekamer, P. L., Deng, X., Mitchell, H. L., Baek, S.-J., & Gagnon, N. (2014). Higher resolution in an operational ensemble Kalman filter. *Monthly Weather Review, 142,* 1143–1162. https://doi.org/10.1175/MWR-D-13-00138.1.

Jeong, J.-H., Linderholm, H. W., Woo, S.-H., Folland, C. K., Kim, B.-M., Kim, S.-J., et al. (2013). Impacts of snow initialization on subseasonal forecasts of surface air temperature for the cold season. *Journal of Climate, 26,* 1956–1972. https://doi.org/10.1175/JCLI-D-12-00159.1.

Koster, R. D., et al. (2010). Contribution of land surface initialization to subseasonal forecast skill: First results from a multi-model experiment. *Geophysical Research Letters, 37,* L02402. https://doi.org/10.1029/2009GL041677.

Lalaurette, F. (2003). Early detection of abnormal weather conditions using a probabilistic extreme forecast index. *Quarterly Journal of the Royal Meteorological Society, 129,* 3037–3057. https://doi.org/10.1256/qj.02.152.

Leutbecher, M., & Palmer, T. N. (2008). Ensemble forecasting. *Journal of Computational Physics, 227,* 3515–3539.

Leutbecher, M., Lock, S.-J., Ollinaho, P., Lang, S. T. K., Balsamo, G., Bechtold, P., et al. (2017). Stochastic representations of model uncertainties at ECMWF: State of the art and future vision. *Quarterly Journal of the Royal Meteorological Society.* https://doi.org/10.1002/qj.3094.

Lin, H., Brunet, G., & Derome, J. (2009). An observed connection between the North Atlantic oscillation and the Madden–Julian oscillation. *Journal of Climate, 22,* 364–380. https://doi.org/10.1175/2008JCLI2515.1.

Magnusson, L., Thorpe, A., Buizza, R., Rabier, F., & Nicolau, J. (2015). Predicting this year's European heat wave. ECMWF. *Newsletter, 145,* 4–5.

Palmer, T. N., & Richardson, D. S. (2014). Decisions, Decisions...! ECMWF. *Newsletter, 141,* 12–13.

Shutts, G. J. (2005). A kinetic energy backscatter algorithm for use in ensemble prediction systems. *Quarterly Journal of the Royal Meteorological Society, 131,* 3079–3102.

Swinbank, R., Kyouda, M., Buchanan, P., Froude, L., Hamill, T. M., Hewson, T. D., et al. (2016). The TIGGE project and its achievements. *Bulletin of the American Meteorological Society, 97,* 49–67.

Toth, Z., & Kalnay, E. (1993). Ensemble forecasting at NMC: The generation of perturbations. *Bulletin of the American Meteorological Society, 74,* 2317–2330.

Toth, Z., & Kalnay, E. (1997). Ensemble Forecasting at NCEP and the breeding method. *Monthly Weather Review, 125,* 3297–3319.

Vitart, F. (2014). Evolution of ECMWF sub-seasonal forecast skill scores. *Quarterly Journal of the Royal Meteorological Society, 140,* 1889–1899. https://doi.org/10.1002/qj.2256.

Vitart, F., Robertson, A. W., & Anderson, D. L. T. (2012). Subseasonal to seasonal prediction project: Bridging the gap between weather and climate. *WMO Bulletin, 61,* 23–28.

Vitart, F., Ardilouze, C., Bonet, A., Brookshaw, A., Chen, M., Codorean, C., et al. (2017). The Subseasonal to Seasonal (S2S) Prediction Project Database. *Bulletin of the American Meteorological Society, 98*, 163–173. https://doi.org/10.1175/BAMS-D-16-0017.1.

Wei, M., Toth, Z., Wobus, R., & Zhu, Y. (2008). Initial perturbations based on the ensemble transform (ET) technique in the NCEP global operational forecast system. *Tellus A, 60*, 62–79.

Yamaguchi, M., & Majumdar, S. J. (2010). Using TIGGE data to diagnose initial perturbations and their growth for tropical cyclone ensemble forecasts. *Monthly Weather Review, 138*, 3634–3655.

Open Access This chapter is distributed under the terms of the Creative Commons Attribution 4.0 International License (http://creativecommons.org/licenses/by/4.0/), which permits use, duplication, adaptation, distribution and reproduction in any medium or format, as long as you give appropriate credit to the original author(s) and the source, a link is provided to the Creative Commons license and any changes made are indicated.

The images or other third party material in this chapter are included in the work's Creative Commons license, unless indicated otherwise in the credit line; if such material is not included in the work's Creative Commons license and the respective action is not permitted by statutory regulation, users will need to obtain permission from the license holder to duplicate, adapt or reproduce the material.

CHAPTER 9

Seasonal-to-Decadal Climate Forecasting

Emma Suckling

Abstract Forecasting climate over the near-term, from a season to decades ahead, has the potential to inform decision-making within the energy sector in a number of ways: from energy trading to scheduling maintenance and resource management. Recent advances in forecasting at these timescales have led to promising levels of skill in predicting the large-scale drivers of seasonal and multi-annual climate variability as well as the consequent local climate impacts of relevance for the energy sector (e.g. seasonal temperatures and wind speeds). This chapter discusses the unique aspects of near-term prediction, how it differs from the task of weather prediction and long-term climate projections, the sources of predictability on these timescales, as well as some of the current climate forecasting tools and products aiming to provide value to the energy sector.

Keywords Seasonal-to-decadal • Forecasting • Climate • Predictability • Skill

E. Suckling (✉)
NCAS-Climate, Department of Meteorology, University of Reading, Reading, UK

© The Author(s) 2018
A. Troccoli (ed.), *Weather & Climate Services for the Energy Industry*,
https://doi.org/10.1007/978-3-319-68418-5_9

Introduction to Climate Forecasting

Climate prediction over the near-term, from seasons to multiple decades ahead, has received much attention over recent years for its potential to inform decisions in areas such as risk management and adaptation planning (Smith et al. 2012; Kirtman et al. 2013; Doblas-Reyes et al. 2013b; Meehl et al. 2014). In theory, local and regional scale forecasts on seasonal-to-decadal timescales could be beneficial to the energy sector, for example in terms of understanding vulnerabilities under different energy mixes, planning future wind farm sites or developing resource management strategies. In practice, such forecasts must demonstrate that they are reliable and to add value to the practices currently adopted for decision-making.

Prediction on seasonal-to-decadal timescales occupies a middle ground between weather forecasting, whose goal is to provide a snapshot of atmospheric conditions at a particular point in time for a few days ahead, and climate projection, which aims to estimate the response to external forcings such as from greenhouse gases and aerosols. The goal of seasonal-to-decadal prediction is generally to provide a statistical summary, or probability distributions, of conditions over the coming months and years given some knowledge of the current climate state, or phase of climate variability. The feasibility of predictions on these timescales arises from components of the climate system that evolve at a slower rate than the atmosphere, such as the ocean and land surface, and the interactions between them (Palmer and Hagedorn 2006; Meehl et al. 2009; Boer 2011). Sources of potential prediction skill include both externally forced low-frequency variability due to anthropogenic factors (such as greenhouse gas and aerosol concentrations and land use changes), as well as natural variations in solar activity and volcanic aerosol, and low-frequency variations within the climate system (such as large-scale modes of variability in the atmosphere and oceans).

Sources of Predictability

On seasonal timescales, the main source of predictability is the coupled ocean-atmosphere El-Niño Southern Oscillation (ENSO) phenomenon (Trenberth et al. 2000; Alexander et al. 2002; Wu et al. 2009), which has been a major factor in the success of seasonal forecasting using both dynamical and statistical models (van Oldenborgh et al. 2005; Coelho et al. 2006; Weisheimer et al. 2009; Wu et al. 2009). This mode of variability in the tropical Pacific is known to influence local-scale seasonal climate in many remote locations over the globe. The Madden Julian oscillation (MJO), in the tropical West Pacific and Indian Oceans, is characterised by an eastward progression of atmospheric convection (Madden and Julian 1971). Studies suggest evidence of teleconnections between the MJO and variability in the extratropics in the Pacific basin (Kim et al. 2006) and Atlantic (Cassou 2008) on monthly timescales. Seasonal predictability is also thought to arise from interactions between the troposphere and stratosphere, associated with the quasi-biennial oscillation (QBO) (Baldwin et al. 2001) and sudden stratospheric warming (SSW) (Marshall and Scaife 2010) which can have an important influence on winter conditions, particularly over Europe (Ineson and Scaife 2009). The North Atlantic Oscillation (NAO) is another source of low-frequency variability often attributed to stratosphere-troposphere coupling (Scaife et al. 2005) and is known to influence winter temperatures and rainfall over Northern Europe and Central Asia (Ineson and Scaife 2009; Matthes et al. 2006). Recent studies have reported skill at predicting winter NAO from seasonal forecasts using dynamical models (Scaife et al. 2014). Evidence of teleconnections between winter NAO and European climate in the following spring has also been suggested based on statistical analyses using observations (Herceg-Bulić and Kucharski 2013). The NAO has also been implicated as a predictor of Northern Hemisphere temperature variability at multidecadal timescales (Li et al. 2013).

On timescales of years to decades ahead, a major source of predictability is likely to arise from slowly evolving (multidecadal) variations in sea surface temperatures (SSTs) in the North Atlantic, referred to as Atlantic Multidecadal Variability (AMV). North Atlantic SST fluctuations are linked to variability in the Atlantic Meridional Overturning Circulation (AMOC), which may vary naturally or through external influences, such as volcanoes or greenhouse gases. It has been suggested that AMV and AMOC could be potentially predictable several years ahead (Griffes and Bryan 1997; Dunstone and Smith 2010) and some evidence suggests that associated changes in climate over Europe, America and the African Sahel, as well as in the strength and position of the Atlantic storm track (Knight et al. 2006; Sutton and Hodson 2007; Sutton and Dong 2012) could also be predictable. Pacific decadal variability (PDV) is also associated with climate impacts over America, Asia, Africa and Australia on multidecadal timescales (Power et al. 1999; Deser et al. 2004); however, the processes involved are currently not well understood and evidence of potential predictability is weaker than for AMV. The Indian Ocean Dipole, characterised as a fluctuation in SSTs, has recently been associated with changes in winds and rainfall across Africa and India on interannual timescales (Webster et al. 1999).

In addition to large-scale modes of internally generated natural variability, the evolution of the climate system is affected by external factors, including a response to anthropogenic activity as well as natural factors such as solar activity and volcanic eruptions. Solar activity is somewhat predictable in terms of sunspot number and solar radiative output, which varies with a period of approximately 11 years. The influence of solar activity on global climate is fairly small and studies suggest global temperatures may vary by about 0.1°C due to solar activity (Lean and Rind 2008), as well as influence stratospheric temperatures and induce small changes in tropical atmospheric circulation (see e.g. Gray et al. 2010, Gray et al. 2016, Lockwood et al. 2010). Volcanic activity cannot be predicted in advance, but large volcanic eruptions have a significant impact on forecast skill once they have occurred. Aerosols emitted into the stratosphere affect global and local temperatures over the timescale of a year or two, as well as the hydrological cycle and atmospheric circulation. Volcanic activity can also impact ocean circulation and heat content on timescales of years or decades, which has important implications for decadal forecast skill (Marshall et al. 2009).

THE PROBABILISTIC NATURE OF CLIMATE FORECASTING

Climate forecasts are necessarily formulated in a probabilistic way (usually as ensembles of deterministic possible outcomes), owing to the inherently uncertain nature of climate prediction. Both the chaotic nature of the climate system (in which small errors in estimating the initial climate state grow with time in any forecast) and inadequacies of the forecast systems themselves (due to approximations in the formulation of the models and missing processes and feedbacks) contribute to forecast uncertainties, which play an important role in the interpretation and use of such forecasts for decision-making. The relative contributions of different sources of forecast uncertainty depend on the timescale, region and variable of interest, but can be difficult to disentangle (see Fig. 9.1). However, at shorter timescales, from months to years ahead, internal variability (from mechanisms such as ENSO or MJO) is typically a major source of uncertainty, accounting for the largest fraction of the variance (e.g. in global or regional temperatures) in studies based on model predictions (Hawkins and Sutton 2009). Over the near-term, up to a few decades ahead, scenario uncertainty (which is assessed through a series of possible pathways that include estimates of future greenhouse gas emissions, land use changes and socio-economic factors) is typically not a dominant source of uncer-

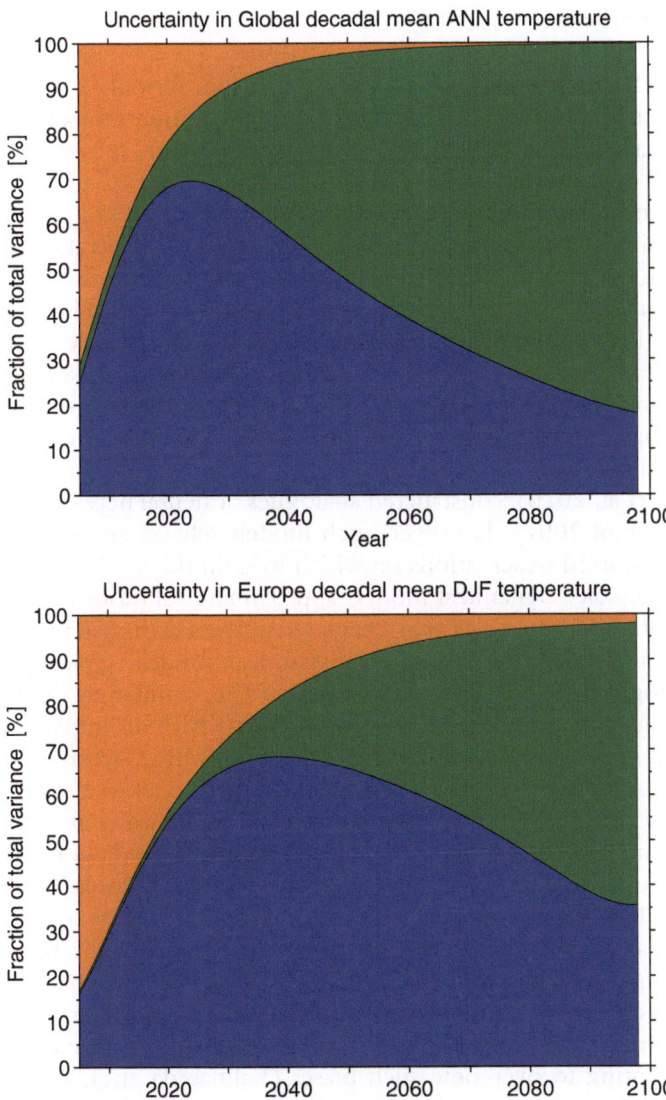

Fig. 9.1 Relative contributions to the fraction of total variance from each source of uncertainty in projections of decadal mean surface air temperature in a) global mean and b) Europe mean. Green regions represent scenario uncertainty, blue regions represent model uncertainty, and orange regions represent the internal variability component. The importance of model uncertainty is clearly visible for all policy-relevant timescales. As the size of the region is reduced, the relative importance of internal variability increases. Scenario uncertainty only becomes important at multidecadal lead times (Hawkins and Sutton 2009, see also Kirtman et al. 2013).

tainty at these timescales and is shown to account for only a small fraction of variance in model studies. Model inadequacy accounts for a large source of uncertainty, particularly in the near-term; therefore, the evaluation and comparison of a variety of forecast approaches and models are crucial both in terms of understanding sources of predictability and skill and in terms of designing and calibrating any prediction system for operational use.

Recent progress in the development of seasonal-to-decadal prediction systems includes the use of empirical methods, dynamical models or a combination of both. Empirical methods are based on observations and exploit statistical relationships to represent physical processes. Empirical models include simple approaches such as characterising the historical climatology, or persisting current conditions, as well as more sophisticated methods such as linear regression models (e.g. Eden et al. 2015; Suckling et al. 2016), constructed analogues or neural networks (see, e.g. van den Dool 2007). However, such models rely on an adequate database of historical observations on which to train the model, which is not always available. Dynamical models approximate solutions to the fundamental physical equations that characterise the Earth system. Seasonal-to-decadal predictions based on dynamical models typically involve combining a boundary condition problem (i.e. simulating the response to forcings and the feedbacks between them) with an initial condition problem, in which the current state of the atmosphere, ocean, cryosphere and land surface is estimated by initialising the model to a state close to observations through data assimilation (see, e.g. Smith et al. 2012). The aim of initialisation is to narrow uncertainty in the predictions by taking into account the phase of internal climate variability (Doblas-Reyes et al. 2013a). The process of initialisation is not trivial, however, and constraining a model with observations generally causes initialisation shocks, which impact the forecast skill. Furthermore, systematic biases cause a model to drift away from the observations over time towards its preferred climatology. Several recent studies have adopted different methodologies for attempting to overcome such biases (Balmaseda et al. 2009); however, at present the common approach to dealing with model bias and drift is to remove any systematic errors through post-processing. Post-processing refers to approaches that are used to transform raw model output into forecast products and includes calibration (e.g. bias adjustment or downscaling) and combination (in which information from different sources and models are combined to form a single forecast) (see, e.g. Doblas-Reyes et al. 2013b).

Assessing the Quality of Climate Forecasts

The quality of seasonal-to-decadal prediction systems is typically assessed by comparing predictions from different models against each other and against observations over a historical hindcast period (Goddard et al. 2013). Hindcasts (sometimes referred to as reforecasts) are essentially retrospective forecasts generated using today's state-of-the-art models, based on knowledge of the historical climate drivers. Several different attributes of the forecast ensemble are often quantified, including model bias and ensemble spread, as well as the correspondence between forecast and outcome pairs using a variety of statistical metrics (Jolliffe and Stephenson 2003). Such metrics include deterministic skill scores, which consider the ensemble mean properties of a set of predictions, and probabilistic skill scores (Bröcker and Smith 2007) that quantify the quality of the full distribution of ensemble members relative to a reference forecast system (such as climatology, persistence or another forecast system) (Suckling and Smith 2013). Reliability measures the correspondence between the predicted probabilities and observed frequencies of a particular set of events. The evaluation of any forecast system in this way is crucial, not only for the development and improvement of systems for operational use, but also in understanding when forecasts are likely to provide reliable information (e.g. Weisheimer and Palmer 2014).

Climate Forecast Tools for the Energy Sector

Both empirical approaches and the use of initialised dynamical prediction systems have been relatively successful for seasonal forecasting, leading to the availability of a number of operational products including the North American Multi-Model Ensemble (NMME)[1] and the EUROSIP multi-model seasonal forecasting system.[2] Initiatives such as the Advancing Renewable Energy with Climate Services (ARECS) and European Climatic Energy Mixes (ECEM)[3] projects aim to develop state-of-the art tools and forecasts that are relevant to the energy sector (see Fig. 9.2).

The field of decadal forecasting is still relatively new, but is rapidly developing. Currently decadal predictions are not widely used as operational products; however, model intercomparison projects such as CMIP5 (Taylor et al. 2012) have advanced the science base for decadal prediction using dynamical model and projects such as the Decadal Forecast Exchange[4] and the World Climate Research Programme (WRCP) Grand Challenge on Near-Term Climate Prediction (https://www.wcrp-climate.org/

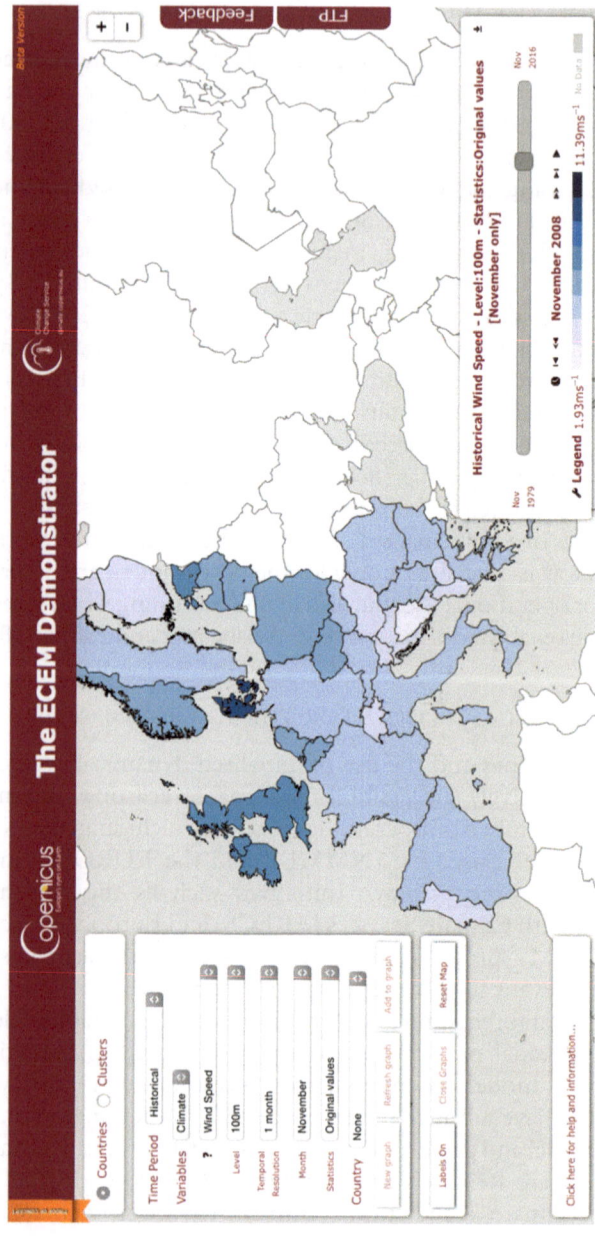

Fig. 9.2 Example of the information available from the ECEM Demonstrator tool (http://ecem.climate.copernicus. eu). Historical monthly mean wind speed for November 1979 over Europe. Essential climate variables and energy impact indicators are available on a range of timescales, including a historical reanalysis, seasonal forecasts and climate projections

grand-challenges/gc-near-term-climate-prediction) aim to facilitate development of decadal forecast approaches towards operational use.

CONCLUDING REMARKS

The recent interest in seasonal-to-decadal forecasting has led to significant improvements in forecast skill, as well as a better understanding of the sources of forecast skill and climate predictability. At seasonal timescales, ENSO is reliably predicted several months ahead by the latest empirical and dynamical models, and ENSO teleconnections are also well predicted (Wang et al. 2009). Recent analyses of the reliability of seasonal forecasts have indicated that temperature predictions in the east and west coast of North America and parts of China and East Asia are reliable, particularly in winter (DJF), while predictions over South America, Southern Africa and Australia are reliable in austral winter (JJA) (see Fig. 9.3) (Weisheimer and Palmer 2014). Predictions in other regions and seasons may also be useful as decision-relevant tools; however, there is often a lot of diversity between different models, making it difficult to make statements about the broad level skill in those regions. Predictions of precipitation are generally less reliable, however. On the other hand, near surface winds are strongly constrained by the ocean in the tropics and are therefore relatively predictable. Furthermore, recent indications of skill in predicting the NAO during winter show promising signals of predictability of winter winds and temperatures in the extratropics, particularly over Europe (Scaife et al. 2014).

On decadal timescales, empirical methods have been demonstrated to represent temperatures well, both for externally forced signals (Lean and Rind 2008; Suckling et al. 2016) and for idealised studies of internal variability (Hawkins et al. 2011). There is also evidence of predictability in north Atlantic SSTs related to skill in predictions of the AMOC (Pohlmann et al. 2013); however, evaluations are currently complicated by a lack of observations over an extended period. The initialisation of model predictions has provided some evidence of improvements in AMV and other large-scale models of variability (Doblas-Reyes et al. 2013a), as well as in predictions of, for example, Atlantic hurricane number more than a year in advance (Smith et al. 2010).

The future direction of climate forecasting for the energy sector will see continued efforts in the development of climate prediction systems, including a new generation of high-resolution global and regional climate models under projects such as PRIMAVERA,[5] as well as improved understanding of the mechanisms that drive local variability and impacts and the

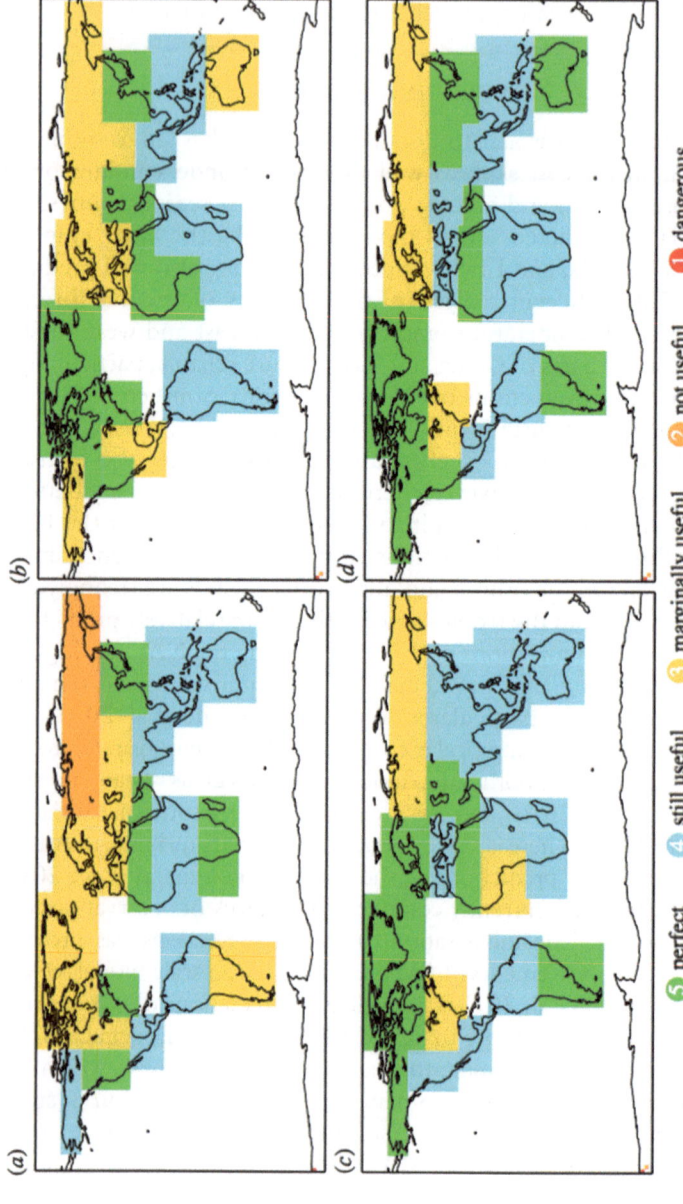

Fig. 9.3 Example of communicating seasonal forecasts skill information. Reliability of the European Centre for Medium-Range Weather Forecasts (ECMWF) Seasonal Forecast System 4 for predictions of 2m temperature during (**a**) cold DJF, (**b**) warm DJF, (**c**) cold JJA and (**d**) warm JJA (Weisheimer and Palmer 2014)

conditions which lead to greater predictability and forecast skill. Initiatives such as the Copernicus Climate Change Service (C3S—climate.copernicus.eu) and its related projects are already working towards tailoring state-of-the-art climate forecasts for the energy sector, including the delivery of energy-relevant climate variables and energy-impact indicators from a season to several decades ahead.

Notes

1. http://www.cpc.ncep.noaa.gov/products/NMME/.
2. https://www.ecmwf.int/en/forecasts/documentation-and-support/long-range/seasonal-forecast-documentation/eurosip-user-guide/multi-model.
3. http://ecem.climate.copernicus.eu.
4. http://www.metoffice.gov.uk/research/climate/seasonal-to-decadal/long-range/decadal-multimodel.
5. https://www.primavera-h2020.eu.

References

Alexander, M. A., et al. (2002). The atmospheric bridge: The influence of ENSO teleconnections on air-sea interactions over the global oceans. *Journal of Climate, 15*, 2205–2231. https://doi.org/10.1175/1520-0442(2002)015.

Baldwin, M. P., et al. (2001). The quasi-biennial oscillation. *Reviews of Geophysics, 39*, 179–229. https://doi.org/10.1029/1999RG000073.

Balmaseda, M., et al. (2009). Impact of initialization strategies and observations on seasonal forecast skill. *Geophysical Research Letters, 36*, L01701. https://doi.org/10.1029/2008GL035561.

Boer, G. J. (2011). Decadal potential predictability of twenty-first century climate. *Climate Dynamics, 36*, 1119–1133. https://doi.org/10.1007/s00382-010-0747-9.

Bröcker, J., & Smith, L. A. S. (2007). Scoring probabilistic forecasts: The importance of being proper. *Weather and Forecasting, 22*, 382–388. https://doi.org/10.1175/WAF966.1.

Cassou, C. (2008). Intraseasonal interaction between the Madden Julian oscillation and the North Atlantic oscillation. *Nature, 455*, 523–527. https://doi.org/10.1038/nature07286.

Coelho, C. A. S., et al. (2006). Towards an integrated seasonal forecasting system for South America. *Journal of Climate, 19*, 3704–3721. https://doi.org/10.1175/JCLI3801.1.

Deser, C., Phillips, A. S., & Hurrell, J. W. (2004). Pacific interdecadal climate variability: Linkages between the tropics and the North Pacific during Boreal

Winter since 1900. *Journal of Climate, 17,* 3109–3124. https://doi.org/10.1175/1520-0442(2004)017<3109:PICVLB>2.0.CO;2.

Doblas-Reyes, F. J., et al. (2013a). Initialized near-term regional climate change prediction. *Nature Communications, 4,* 1715. https://doi.org/10.1038/ncomms2704.

Doblas-Reyes, F. J., et al. (2013b). Seasonal climate predictability and forecasting: Status and prospects. *WIREs: Climate Change, 4,* 245–268. https://doi.org/10.1002/wcc.217.

van den Dool, H. M. (2007) *Empirical methods in short-term climate prediction,* Oxford University Press, Oxford. ISBN:0-19-920278-8.

Dunstone, N. J., & Smith, D. M. (2010). Impact of atmosphere and sub-surface ocean data on decadal climate prediction. *Geophysical Research Letters, 37,* L02709. https://doi.org/10.1029/2009GL041609.

Eden, J. M., et al. (2015). A global empirical system for probabilistic seasonal climate prediction. *Geoscientific Model Development Discussion, 8,* 3941–3970. https://doi.org/10.5194/gmdd-8-3941-2015.

Goddard, L., et al. (2013). A verification framework for interannual-to-decadal prediction experiments. *Climate Dynamics, 40,* 245–272. https://doi.org/10.1007/s00382-012-1481-2.

Griffes, S. M., & Bryan, K. (1997). Predictability of North Atlantic multidecadal climate variability. *Science, 275*(5297), 181–184. https://doi.org/10.1126/science.275.5297.181.

Hawkins, E., & Sutton, R. (2009). The potential to narrow uncertainty in regional climate predictions. *Bulletin of the American Meteorological Society, 90,* 1095. https://doi.org/10.1175/2009BAMS2607.1.

Hawkins, E., et al. (2011). Evaluating the potential for statistical decadal predictions of sea surface temperatures with a perfect model approach. *Climate Dynamics, 37,* 2495–2509. https://doi.org/10.1007/s00382-011-1023-3.

Herceg-Bulić, I., & Kucharski, F. (2013). North Atlantic SSTs as a link between the wintertime NAO and the following spring climate. *Journal of Climate, 27,* 186. https://doi.org/10.1175/JCLI-D-12-00273.1.

Ineson, S., & Scaife, A. A. (2009). The role of the stratosphere in the European climate response to El Niño. *Nature Geoscience, 2,* 32–36. https://doi.org/10.1038/ngeo381.

Jolliffe, I. T. and Stephenson, D. B. (2003) *Forecast verification: A practitioner's guide in atmospheric science,* Chichester; West Sussex: J. Wiley. ISBN:978-0-470-66071-3.

Kim, B.-M., Lim, B.-H., & Kim, K.-Y. (2006). A new look at the midlatitude-MJO teleconnection in the Northern Hemisphere Winter. *Quarterly Journal of the Royal Meteorological Society, 132,* 485–503. https://doi.org/10.1256/qj.04.87.

Kirtman, B., et al. (2013). Near-term climate change: Projections and predictability. In T. F. Stocker, D. Qin, G.-K. Plattner, M. Tignor, S. K. Allen, J. Boschung,

A. Nauels, Y. Xia, V. Bex, & P. M. Midgley (Eds.), *Climate change 2013: The physical science basis. Contribution of working group I to the fifth assessment report of the intergovernmental panel on climate change* (pp. 953–1028). Cambridge; New York: Cambridge University Press. https://doi.org/10.1017/CB09781107415324.023.

Knight, J. R., Folland, C. K., & Scaife, A. A. (2006). Climate impacts of the Atlantic Multidecadal Oscillation. *Geophysical Research Letters, 33*, L17706. https://doi.org/10.1029/2006GL026242.

Lean, J. L., & Rind, D. H. (2008). How natural and anthropogenic influences alter global and regional surface temperatures: 1889 to 2006. *Geophysical Research Letters, 35*, L18701. https://doi.org/10.1029/2008GL034864.

Li, J., Sun, C., & Jin, F.-F. (2013). NAO implicated as a predictor of Northern Hemisphere mean temperature multidecadal variability. *Geophysical Research Letters, 40*, 5497–5502. https://doi.org/10.1002/2013GL057877.

Madden, R. A., & Julian, P. R. (1971). Detection of a 40–50 day oscillation in the zonal wind in the tropical Pacific. *Journal of the Atmospheric Sciences, 28*, 702–708. https://doi.org/10.1175/1520-0469(1971)028<0702:DOADOI>2.0.CO;2.

Marshall, A. G., Scaife, A. A., & Ineson, S. (2009). Enhanced seasonal prediction of European winter warming following volcanic eruptions. *Journal of Climate, 22*, 6168–6180. https://doi.org/10.1175/2009JCLI3145.1.

Marshall, A. G., & Scaife, A. A. (2010). Improved predictability of stratospheric sudden warming events in an atmospheric general circulation model with enhanced stratospheric resolution. *Journal of Geophysical Research, 115*, D16114. https://doi.org/10.1029/2009JD012643.

Matthes, K., et al. (2006). Transfer of the solar signal from the stratosphere to the troposphere: Northern winter. *Journal of Geophysical Research, 111*, D06108. https://doi.org/10.1029/2005JD006283.

Meehl, G. A., et al. (2009). Decadal prediction: Can it be skillful? *Bulletin American Meteorological Society, 90*, 1467–1485. https://doi.org/10.1175/2009BAMS2778.1.

Meehl, G. A., et al. (2014). Decadal climate prediction: An update from the trenches. *Bulletin of the American Meteorological Society, 95*, 2. https://doi.org/10.1175/BAMS-D-12-00241.1.

van Oldenborgh, G.-J., et al. (2005). Did the ECMWF seasonal forecast model outperform statistical ENSO forecast models over the last 15 years? *Journal of Climate, 18*, 3240–3249. https://doi.org/10.1175/JCLI3420.1.

Palmer, T., & Hagedorn, R. (Eds.). (2006). *Predictability of weather and climate.* Cambridge: Cambridge University Press. https://doi.org/10.1017/CB09780511617652.

Pohlmann, H., et al. (2013). Skillful predictions of the mid-latitude Atlantic meridional overturning circulation in a multi-model system. *Climate Dynamics, 41*, 775–785. https://doi.org/10.1007/s00382-013-1663-6.

Power, S., et al. (1999). Interdecadal modulation of the impact of ENSO on Australia. *Climate Dynamics, 15,* 319–324. https://doi.org/10.1007/s003820050284.

Scaife, A. A., et al. (2005). A stratospheric influence on the winter NAO and North Atlantic surface climate. *Geophysical Research Letters, 32,* L18715. https://doi.org/10.1029/2005GL023226.

Scaife, A. A., et al. (2014). Skilful long range prediction of European and North American winters. *Geophysical Research Letters, 41,* 2514–2519. https://doi.org/10.1002/2014GL059637.

Smith, D. M., et al. (2010). Skilful multi-year predictions of Atlantic hurricane frequency. *Nature Geoscience, 3,* 846–849. https://doi.org/10.1038/ngeo1004.

Smith, D. M., Scaife, A. A., & Kirtman, B. P. (2012). What is the current state of scientific knowledge with regard to seasonal and decadal forecasting? *Environmental Research Letters, 7,* 015602. https://doi.org/10.1088/1748-9326/7/1/015602.

Suckling, E. B., & Smith, L. A. (2013). An evaluation of decadal probability forecasts from state-of-the-art climate models. *Journal of Climate, 26,* 9334–9347. https://doi.org/10.1175/JCLI-D-12-00485.1.

Suckling, E. B., van Oldenborgh, G.-J., Eden, J. M., & Hawkins, E. (2016). An empirical model for probabilistic decadal prediction: Global attribution and regional hindcasts. *Climate Dynamics.* https://doi.org/10.1007/s00382-016-3255-8.

Sutton, R., & Hodson, D. (2007). Climate response to basin-scale warming and cooling of the North Atlantic Ocean. *Journal of Climate, 20*(5), 891–907. https://doi.org/10.1175/JCLI4038.1.

Sutton, R., & Dong, B. (2012). Atlantic Ocean influence on a shift in European climate in the 1990s. *Nature Geoscience, 5,* 788–792. https://doi.org/10.1038/ngeo1595.

Trenberth, K. E., et al. (2000). The Southern Oscillation revisited: Sea level pressures, surface temperatures and precipitation. *Journal of Climate, 13,* 4358–4365. https://doi.org/10.1175/1520-0442(2000)013.

Wang, B., et al. (2009). Advance and prospectus of seasonal prediction: Assessment of the APCC/CliPAS 14-model ensemble retrospective seasonal prediction (1980–2004). *Climate Dynamics, 33,* 93–117. https://doi.org/10.1007/s00382-008-0460-0.

Webster, P. J., et al. (1999). Coupled ocean-atmosphere dynamics in the Indian Ocean during 1997–98. *Nature, 401,* 356–360. https://doi.org/10.1038/43848.

Weisheimer, A., et al. (2009). ENSEMBLES – A new multi-model ensemble for seasonal-to-annual predictions: Skill and progress beyond DEMETER in forecasting tropical Pacific SSTs. *Geophysical Research Letters, 36,* L21711. https://doi.org/10.1029/2008GL040896.

Weisheimer, A., & Palmer, T. (2014). On the reliability of seasonal climate fore-casts. *Journal of the Royal Society Interface, 11*(162), 20131. https://doi.org/10.1098/rsif.2013.1162.

Wu, R., Kirtman, B. P., & van den Dool, H. (2009). An analysis of ENSO predic-tion skill in the CFS retrospective forecasts. *Journal of Climate, 22,* 1801–1818. https://doi.org/10.1175/2008JCLI2565.1.

Open Access This chapter is distributed under the terms of the Creative Commons Attribution 4.0 International License (http://creativecommons.org/licenses/by/4.0/), which permits use, duplication, adaptation, distribution and reproduc-tion in any medium or format, as long as you give appropriate credit to the original author(s) and the source, a link is provided to the Creative Commons license and any changes made are indicated.

The images or other third party material in this chapter are included in the work's Creative Commons license, unless indicated otherwise in the credit line; if such material is not included in the work's Creative Commons license and the respective action is not permitted by statutory regulation, users will need to obtain permission from the license holder to duplicate, adapt or reproduce the material.

CHAPTER 10

Regional Climate Projections

Robert Vautard

Abstract When designing adaptation and mitigation measures of climate change for the coming decades and up to the middle of the century, policymakers and industries must rely upon climate information that is at an appropriate scale to evaluate impacts, vulnerabilities and risks due to changes in climate. It is, therefore, essential that the quantitative information on the climate and its impacts is reliable. Reliable quantitative information about climate change impacts must also be available. This includes estimations of uncertainty bounds. In the current state of knowledge, technology and structure of scientific communities, climate change impact studies are achieved from a suite of models: global earth system models, with a generally low-resolution (100–300 km), regional limited-area climate models with a higher resolution (10–50 km), which take their boundary conditions from global models and impact models calculating how changes in weather, ocean and biogeochemical cycles affect the system to be adapted.

Keywords Downscaling • Uncertainty • Extreme events • Climate projections • Scenarios

R. Vautard (✉)
Laboratoire des Sciences du Climat et de l'Environnement,
Institut Pierre-Simon Laplace, Paris, France

© The Author(s) 2018   139
A. Troccoli (ed.), *Weather & Climate Services for the Energy Industry*,
https://doi.org/10.1007/978-3-319-68418-5_10

Regional climate projections

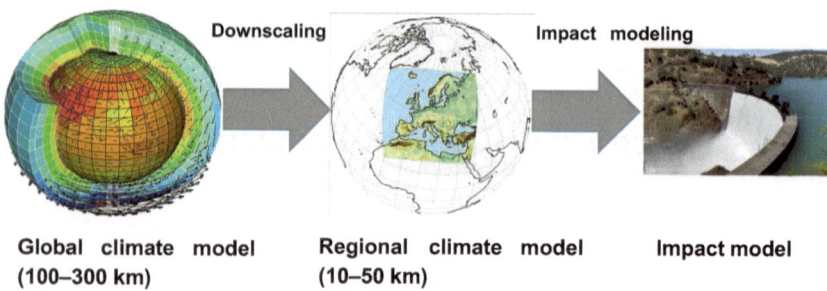

| Downscaling | Impact modeling |

| Global climate model (100–300 km) | Regional climate model (10–50 km) | Impact model |

Fig. 10.1 Schematic of the modelling chain used to calculate the impacts of climate change. In this illustration, the impacts can be the river discharge or hydropower potential

INTRODUCTION

Regional climate projections are the key chain elements that provide information at a scale allowing impacts calculations, and the assessment of adaptation measures. Regional climate projections also allow an improved simulation of physical and biogeochemical processes, and of extreme events at the scales relevant for impacts studies (see schematic of Fig. 10.1).

Uncertainty in climate projections cannot be directly estimated by comparing them against observations, as in the case of weather forecasts, due to the long timescales involved. This is why uncertainty is generally estimated from the spread of ensembles of simulations. The larger the spread, the larger the uncertainty. This emphasizes the need to estimate impacts of climate change using ensembles of models. This section describes (1) the specific nature of climate projections as compared to forecasts previously presented, their main underlying assumptions, their uncertainty sources and how the science community is organized; (2) how regional climate projections add value to the global climate projections and (3) how they can be used in the energy sector.

WHAT ARE CLIMATE PROJECTION AND HOW DO THEY DIFFER FROM WEATHER FORECASTS AND DECADAL PREDICTIONS?

Climate projections aim at predicting future climatologies, that is, the statistics of weather, of the state of the ocean, cryosphere, vegetation and atmospheric chemical composition. They are based on a number of assumptions about human evolution (population, economy, land use, land management, technologies and climate policies) and our understanding of the earth systems response to this evolution. Projections are considered as a "boundary value problem", where climate responds to external drivers. The principle is very different from that of weather forecasts, which is an initial value problem whose aim is to provide a snapshot of the state of the atmosphere at particular points in space and time (see also Chap. 6). Climate projections also differ from weather forecasts in that they use fully fledged earth system models describing the evolution of as many as possible variables concerning all compartments of the earth envelope, for example the ocean, cryosphere, as well as the atmosphere. Weather forecasts, when, for instance, limited to a few weeks, do not need to include changes in the oceans as well as other slowly evolving components of the earth system.

Climate projections also aim at explaining the evolution of past climate, from the instrumental period to paleoclimatic periods. A correct simulation of past periods, including the last century, millennium and beyond (Braconnot et al. 2012) provides some confidence to the models ability to simulate climate change, even though comparison with paleo observations often faces numerous scientific challenges. However, as demonstrated in the fifth assessment report of the IPCC WGI (Intergovernmental Panel on Climate Change Working Group I) (IPCC WGI 2013), climate evolutions along the twentieth century, and in particular the late warming, attributed to human influence on atmospheric composition, are fairly well reproduced: the amplitude of the warming is reproduced as well as the long-term modulations, thought to be due to aerosols.

Climate projections used for adaptation are usually initialized from an equilibrium state in a period when human influence is assumed to be minor relative to external natural forcing (solar and volcanic), in the middle of the nineteenth century. Since the memory of initial conditions disappears quickly, it is therefore not expected that projections synchronously represent the actual chaotic short-term observed evolutions across the instrumental period. Instead, one expects climate projections to represent possible trajectories of the weather, and the probability distribution functions (PDFs) of these fluctuations.

However, models have biases difficult to fully control, because of accumulation of approximations, which are due to low-resolution and insufficiently well-represented physical or biogeochemical processes such as convection or land-atmosphere interactions. In models, all these processes are in a balance that is typically shifted as compared to the real world. However, it is often assumed that such biases do not hinder a correct simulation of the response to external forcing. These potential error sources are often called "structural uncertainties".

Two other major sources of uncertainties are also present: the evolution of societies, climate policies and their effect on resulting atmospheric composition (greenhouse gases, aerosols), and on land use. Strong mitigation scenarios, including potential use of negative emission technologies, are likely to significantly alter land use (Smith et al. 2016) and therefore impact climate through an alternative pathway from atmospheric composition. Climate evolutions also undergo a natural internal variability at a timescale of a decade to several decades. Multidecadal modes and oscillations have long been identified (Ghil and Vautard 1991), with yet no full understanding. Such modes can induce temporary warming hiatus such as the one witnessed in the last decade or so. It is considered that the latter uncertainty source dominates for the few coming decades, while uncertainty on scenarios is the dominant driver for the second half of the century.

The global climate projections production that feed the IPCC assessment reports are internationally coordinated by the Climate Model Intercomparison Project (CMIP) (Taylor et al. 2012) supported by the World Climate Research Program. Modelling centres from around the world carry out a core set of simulation using the same natural forcings and socio-economic scenario assumptions. This makes simulations comparable and allows investigation of spread across the different models and some sources of uncertainty to be estimated.

International coordination not only takes place for climate simulations but also for data dissemination through the distributed Earth System Grid Federation (ESGF), with common standards, vocabulary and quality control. The ESGF data nodes[1], where climate simulations are available from, are part of a gigantic internationally coordinated climate data repository, with a number of functionalities (dynamic catalogue, quick look visualizations etc.). This coordinated effort allows for a systematic retrieval and use of model ensembles in order to account for climate projection uncertainties. The archive is now open to data from a number of projects, namely not just CMIP.

Regional Climate Projections

After a few pioneering projects such as PRUDENCE (Christensen and Christensen 2007), ENSEMBLES (Hewitt and Griggs 2004) or the North American Regional Climate Change Assessment Program (NARCCAP) (Mearns et al. 2012), regional climate projections were only recently coordinated at an international scale within the framework of the COordinated Regional Downscaling Experiment (CORDEX) (Giorgi et al. 2009). This framework aims at providing downscaled global climate projections through a coordinated approach over several regions of the world, using refined, but scale-limited models which take global model output as boundary conditions. A number of regions of the globe are now modelled using this approach, including Europe (Jacob et al. 2014). In Europe, these coordinated projections were made both at a low (50 km) and high (12 km) resolutions.

Regional climate projections are issued from limited-area regional climate models (RCMs) (Giorgi and Mearns 1999) which describe weather in a limited area (typically a region of around 25 thousand of kilometres square) with a higher resolution (typically 10–50 km) than global climate models (GCMs). These models are often only atmospheric models, forced by sea-surface temperatures and lateral boundary values of temperature, humidity and wind prescribed to be the values obtained from GCMs. Therefore, regional models solve the atmospheric equations with constraints on the boundaries. Recently, regional earth system models are also currently being developed, using ocean coupled with the atmosphere, which is a central development of the MED-CORDEX[2] project (Ruti et al. 2015).

Just as is true for GCMs, RCMs have inherent climate biases that can be quite large when underpinning processes are poorly constrained by observations. Such is, for instance, the case of heat waves in Europe (Vautard et al. 2013). These biases add to those already present in boundary conditions provided by the GCMs. In particular, providing climate projections at a higher resolution does not annihilate potential biases in large-scale dynamical structures (jet streams, weather patterns), inherited from GCMs.

The expectation, however, is that RCMs better simulate atmospheric flows, and therefore weather, in the vicinity of marked topography or coastal areas, and other small-scale phenomena, compared to GCMs. Heavy precipitations are usually small-scale phenomena, and recent studies show a clear improvement in the statistics of such phenomena (Prein et al. 2016; Giorgi et al. 2016), in particular over mountainous areas, when using high-resolution simulations (e.g. 12 km as in EURO-CORDEX). This has a clear benefit for use in climate projections for studying applications such as winter tourism or mountain ecosystem impacts. By contrast, no added value of higher resolution was found for larger-scale phenomena such as heat waves (Vautard et al. 2013).

As for GCMs, ensembles of RCMs are assumed to provide a picture of uncertainties in projections. In order to cover a wider range of possibilities, a full GCM-RCM matrix should in principle be explored, each RCM downscaling all GCMs. In practice, while some strategies to explore all combinations have been developed (Mearns et al. 2012), these have not been implemented in practice. As of 2016, the EURO-CORDEX matrix largely remains to be filled for the high-resolution simulations with a 12×12 km^2 grid.

Regional climate projections have climatological biases that usually need to be accounted for in impact studies. Bias adjustment is a statistical process that modifies model simulation values in order to adjust distributions to observed values. It uses a range of methods from simple mean bias removal to quantile mapping, in order to adjust the full distribution. More sophisticated schemes account both for past climate corrections and the evolution of distributions as simulated in models, such as the Cumulative Distribution Function transform (CDFt) method (Vrac et al. 2012; Vrac et al. 2016). It is an assumption that bias adjustment does not deteriorate the climate change signal, or even improves projections of changes. However, it is difficult to prove this in practice. It is therefore recommended that climate projections be processed by several bias adjustment methods.

The following example (see data description below) uses the CDFt method to bias adjust the number of rain days according to recent improvements (Vrac et al. 2016) using ten EURO-CORDEX simulations (available from the ESGF archive). Figure 10.2 shows an example of multi-model mean change in daily precipitation amount between the last 30 years of the twenty-first century and the corresponding period of the twentieth century. As an example, a clear, robust signal of precipitation increase is found in Northern Europe, and of precipitation decrease in Southern Europe, where more than eight models agree on the sign of the change.

THE USE OF CLIMATE PROJECTIONS FOR THE ENERGY SECTOR

Regional climate projections can be used in several ways to help the energy industry and policymakers. Energy demand, renewable energy resources, risks from extreme events, cooling water for thermoelectric generation and other operation conditions all depend on weather and climate. The exposure to weather and climate variability and change will increase in the future decades owing to the tremendous energy transition that is

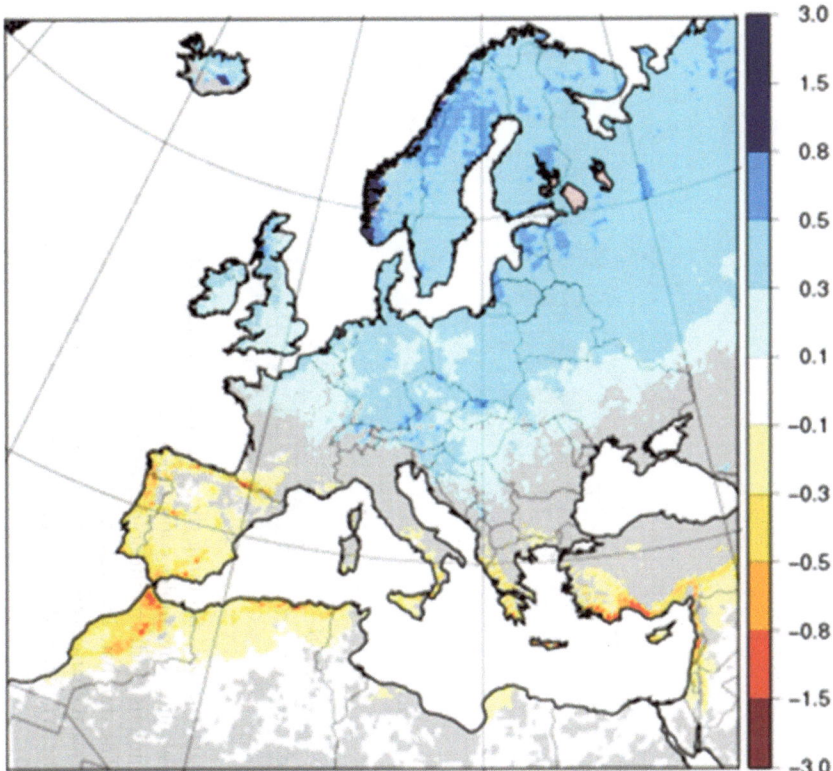

Fig. 10.2 Mean changes in daily precipitation amounts estimated from ten EURO-CORDEX high-resolution model simulations (Jacob et al. 2014), in the RCP8.5 scenario. Changes are measured as differences of mean values calculated over the last 30 years of the twenty-first and twentieth centuries, averaged over the ten projections. Change values, represented by coloured areas, are only displayed when nine or ten models agree on the sign of change. When not, the area is coloured with grey

required to almost completely decarbonize the electricity generation by 2050 in order to reach ambitious climate targets (IPCC WGII 2014).

Both hydropower and thermoelectric generation could be at risk with climate change (van Vliet et al. 2016). In Europe, climate change may be expected to induce rather general decreases in wind power (Tobin et al. 2016), with a higher and more significant signal in Southern Europe (see Fig. 10.3, drawn from the results of the FP7 IMPACT2C project, WP6).

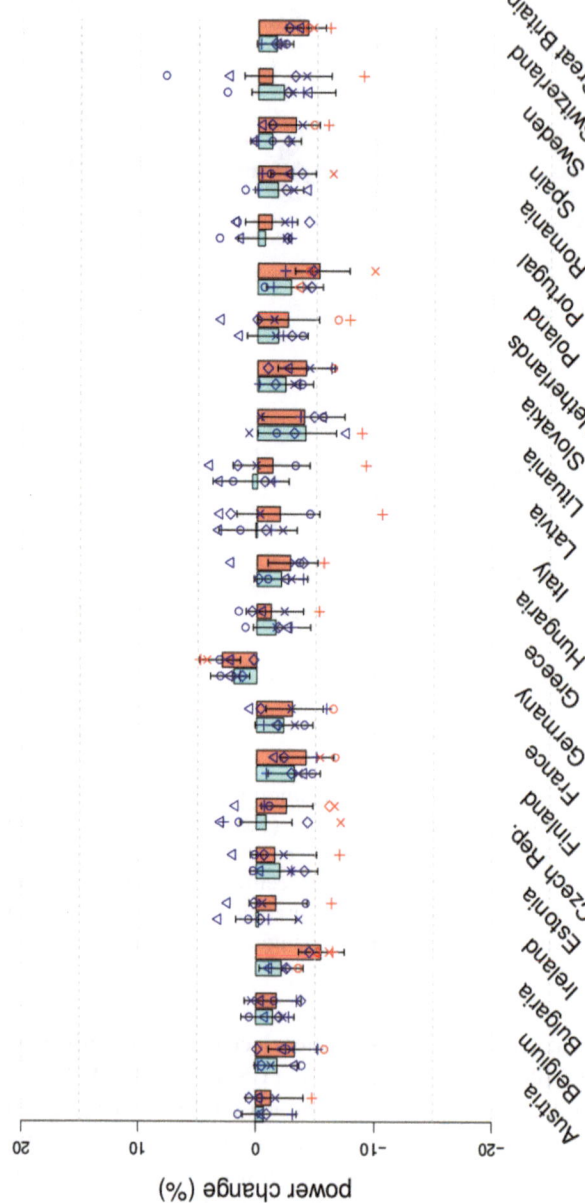

Fig. 10.3 Changes in mean wind power capacity factor, assuming the installed national wind farms fleets as of 2013, under a 2°C (cyan) and 3°C (salmon) global warming. Ensembles consist of 5 simulations out of the 9 RCP4.5-RCP8.5 simulations that reach 2°C or 3°C global warming, using the methodology defined in Vautard et al. (2014). The coloured wide bars indicate the model ensemble mean. The thin bars indicate the 95% level confidence interval as computed using the Wilcoxon-Mann-Whitney test. Model individual changes are represented by differing symbols: symbols are red when changes are significant at the 95% level using the Wilcoxon-Mann-Whitney test (adapted from Tobin et al. 2016 and the results from the FP7 IMPACT2C project, WP6)

Solar photovoltaic (PV) power may also be expected to decrease in Europe as a result of the analysis of an ensemble of regional climate projections (Jerez et al. 2015).

Climate projections, generally available at the daily timescale, and in some cases at sub-daily timescale, can also be used as "weather generators". Bias-adjusted time series obtained from ensembles of simulations are usually 10–30 times longer than observed time series, due to the number of models available and do not suffer from homogeneity problems. They can generate statistics dedicated to the user's problem, such as the risk of typical extreme events. In some cases, infrastructures are built to cope with reference events, such as, for instance, extreme cold spells (e.g. winter of 1963) that occurred in the past. The risk of such events in current and future climates can be evaluated from the climate projection ensembles time series, although this is still a developing science. Reference events and their odds in the current and future climate can provide concrete information for adaptation.

For the needs of the energy sector, dedicated regional climate projections data sets have been developed and made available through prototype "climate services", such as, for Europe, CLIM4ENERGY[3] and European Climatic Energy Mixes (ECEM)[4] Copernicus Climate Change Service projects. Other data sets for other regions (e.g. NARCCAP data for North America[5]) are also available to be used for application in the energy sector. The value of such datasets for adaptation needs is still to be assessed by users as their use is relatively new.

Notes

1. See e.g. http://esgf-node.ipsl.upmc.fr/.
2. MED-CORDEX is the name of the experiment, and stands for Mediterranean.
3. http://clim4energy.climate.copernicus.eu
4. http://ecem.climate.copernicus.eu
5. http://www.narccap.ucar.edu/

References

Braconnot, P., Harrison, S. P., Kageyama, M., Bartlein, P. J., Masson-Delmotte, V., Abe-Ouchi, A., et al. (2012). Evaluation of climate models using palaeoclimatic data. *Nature Climate Change, 2*(6), 417–424.

Christensen, J. H., & Christensen, O. B. (2007). A summary of the PRUDENCE model projections of changes in European climate by the end of this century. *Climatic change, 81*(1), 7–30.

Ghil, M., & Vautard, R. (1991). Interdecadal oscillations and the warming trend in global temperature time series. *Nature, 350,* 324–327.

Giorgi, F., & Mearns, L. O. (1999). Introduction to special section: Regional climate modeling revisited. *Journal of Geophysical Research: Atmospheres, 104*(D6), 6335–6352.

Giorgi, F., Jones, C., & Asrar, G. R. (2009). Addressing climate information needs at the regional level: The CORDEX framework. *World Meteorological Organization (WMO) Bulletin, 58*(3), 175.

Giorgi, F., Torma, C., Coppola, E., Ban, N., Schär, C., & Somot, S. (2016). Enhanced summer convective rainfall at Alpine high elevations in response to climate warming. *Nature Geoscience.* https://doi.org/10.1038/ngeo2761.

Hewitt, C. D., & Griggs, D. J. (2004). Ensembles-based predictions of climate changes and their impacts (ENSEMBLES). *Eos, 85*(52), 566.

IPCC. (2013). *Climate change 2013: The physical science basis* (T. F. Stocker et al., Eds.). New York: Cambridge University Press.

IPCC, WGII: Mach, K., & Mastrandrea, M. (2014). *Climate change 2014: Impacts, adaptation, and vulnerability* (Vol. 1). C. B. Field & V. R. Barros (Eds.). Cambridge and New York: Cambridge University Press.

Jacob, D., Petersen, J., Eggert, B., Alias, A., Christensen, O. B., Bouwer, L. M., et al. (2014). EURO-CORDEX: New high-resolution climate change projections for European impact research. *Regional Environmental Change, 14,* 563–578.

Jerez, S., Tobin, I., Vautard, R., Montávez, J. P., López-Romero, J. M., Thais, F., et al. (2015). The impact of climate change on photovoltaic power generation in Europe. *Nature Communications.* https://doi.org/10.1038/ncomms10014.

Mearns, L. O., Arritt, R., Biner, S., Bukovsky, M. S., McGinnis, S., Sain, S., et al. (2012). The North American regional climate change assessment program: Overview of phase I results. *Bulletin of the American Meteorological Society, 93*(9), 1337–1362.

Prein, A., Gobiet, A., Truehetz, H., Keuler, K., Goergen, K., Teichmann, C., et al. (2016). Precipitation in the EURO-CORDEX 0.11° and 0.44° simulations: High resolution, high benefits? *Climate Dynamics, 46*(1–2), 383–412.

Ruti, P., Somot, S., Dubois, C., Calmanti, S., Ahrens, B., Alias, A., et al. (2015). MED-CORDEX initiative for Mediterranean climate studies. *Bulletin of the American Meteorological Society, 97,* 1175.

Smith, P., Davis, S. J., Creutzig, F., Fuss, S., Minx, J., Gabrielle, B., et al. (2016). Biophysical and economic limits to negative CO_2 emissions. *Nature Climate Change, 6*(1), 42–50.

Taylor, K. E., Stouffer, R. J., & Meehl, G. A. (2012). An overview of CMIP5 and the experiment design. *Bulletin of the American Meteorological Society, 93*(4), 485–498.

Tobin, I., Jerez, S., Vautard, R., Thais, F., Déqué, M., Kotlarski, S., et al. (2016). Climate change impacts on the power generation potential of a European mid-century wind farms scenario. *Environ. Res. Lett.,* *11*(3), 034013.

Vautard, R., Gobiet, A., Jacob, D., Belda, M., Colette, A., Déqué, M., et al. (2013). The simulation of European heat waves from an ensemble of regional climate models within the EURO-CORDEX project. *Climate Dynamics, 41,* 2555–2575.

Vautard, R., Gobiet, A., Sobolowski, S., Kjellström, E., Stegehuis, A., Watkiss, P., et al. (2014). The European climate under a 2 °C global warming. *Environmental Research Letters.* https://doi.org/10.1088/1748-9326/9/3/034006.

van Vliet, M. T. H., Wiberg, D., Leduc, S., & Riahi, K. (2016). Power-generation system vulnerability and adaptation to changes in climate and water resources. *Nature Climate Change, 6*(4), 375–380.

Vrac, M., Drobinski, P., Merlo, A., Herrmann, M., Lavaysse, C., Li, L., et al. (2012). Dynamical and statistical downscaling of the French Mediterranean climate: Uncertainty assessment. *Natural Hazards and Earth System Sciences, 12,* 2769–2784.

Vrac, M., Noël, T., & Vautard, R. (2016). Bias correction of precipitation through Singularity Stochastic Removal: Because occurrences matter. *Journal of Geophysical Research, 121,* 5237–5258.

Open Access This chapter is distributed under the terms of the Creative Commons Attribution 4.0 International License (http://creativecommons.org/licenses/by/4.0/), which permits use, duplication, adaptation, distribution and reproduction in any medium or format, as long as you give appropriate credit to the original author(s) and the source, a link is provided to the Creative Commons license and any changes made are indicated.

The images or other third party material in this chapter are included in the work's Creative Commons license, unless indicated otherwise in the credit line; if such material is not included in the work's Creative Commons license and the respective action is not permitted by statutory regulation, users will need to obtain permission from the license holder to duplicate, adapt or reproduce the material.

CHAPTER 11

The Nature of Weather and Climate Impacts in the Energy Sector

David J. Brayshaw

Abstract The power sector's meteorological information needs are diverse and cover many different distinct applications and users. Recognising this diversity, it is important to understand the general nature of how weather and climate influence the energy sector and the implications they have for quantitative impact modelling. Using conceptual examples and illustrations from recent research, this chapter argues that the traditional 'transfer function' approach that is common to many industrial applications of weather and climate science—whereby weather can be directly mapped to an energy impact—is inadequate for many important power system applications (such as price forecasting and system operations and planning). The chapter concludes by arguing that a deeper understanding of how meteorological impacts in the energy sector are modelled is required.

Keywords Climate variability • Energy sector • Power • Impact modelling • Weather impact • Meteorological information • Power demand

D.J. Brayshaw (✉)
Department of Meteorology, University of Reading, Reading, UK

National Centre for Atmospheric Science, Reading, UK

© The Author(s) 2018
A. Troccoli (ed.), *Weather & Climate Services for the Energy Industry*,
https://doi.org/10.1007/978-3-319-68418-5_11

151

WEATHER AND CLIMATE IMPACTS IN THE ENERGY SECTOR

The power sector's meteorological information needs are diverse. On the one hand, Transmission System Operators (TSOs) may be concerned with detailed geographical forecasts of wind power and demand at relatively short lead times (hours or days ahead) for the operational management of the power grid. This contrasts, for example, with long-term investors in infrastructure and system planners who require a longer view of system resilience (years to decades), and energy traders or maintenance planners seeking to position themselves for the coming weeks or seasons.

A common theme, however, is the need for a series of conversions to transform meteorological information into an actionable decision. The three steps in Fig. 11.1 can typically be recognised.

Chapter 6 discussed the first step in this process at length, and understanding user needs and preferences is discussed elsewhere in other chapters of this book (Chaps. 1, 3, 4 and 5). Here, the focus is on the general process of modelling energy system impacts using meteorological data from numerical simulations, illustrated with selected examples (i.e., Impact Simulation). It is, however, noted that user preferences—once elicited and expressed quantitatively—can be thought of as a conversion of a physical impact (in terms of MWh, prices, loss of load) into a 'utility' (a numerical expression of the user's preferences). To some extent, they can therefore be considered as direct extensions of the impact models discussed below.

It is helpful to identify three distinct levels of complexity in weather- and climate-impact modelling, as illustrated in Fig. 11.2.

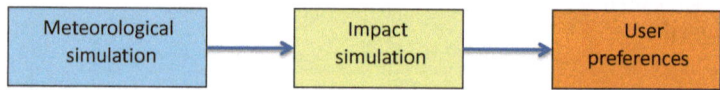

Fig. 11.1 The process of converting meteorological data into actionable information

Fig. 11.2 Levels of impact complexity

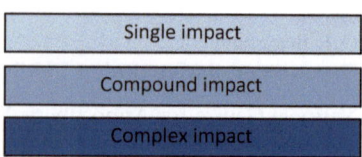

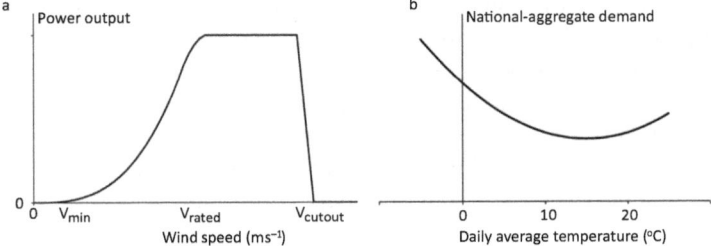

Fig. 11.3 Simple examples of idealised transfer functions used to convert meteo-rological quantities into estimates of power system properties: (**a**) an idealised wind power curve based on Brayshaw et al. (2011); (**b**) a simplified demand model based on Bloomfield et al. (2016). In each example here, the transfer function is shown to depend only on a single meteorological variable for simplicity but in general they may incorporate many input variables. Additional dependencies may be meteorological (e.g., wind direction for wind power, cloud cover for demand) or non-meteorological (e.g., day-of-week for demand), and include stochastic 'noise' to simulate the error and uncertainty in the transfer function

The simplest level, commonly referred to as 'point forecasting', may be defined as the response of a *single energy system component* to a set of mete-orological drivers for which a *transfer function* can be written. Typical examples might include predicting wind power output for a particular tur-bine, farm or country, or forecasting power demand over a particular geo-graphical region (Fig. 11.3). The key aspect is that it is possible to write (or otherwise estimate in at least an approximate form) a function, f, which converts a set of meteorological variables, $\{m\}$, into the energy system property of interest E:

$$E = f\left(\{m\}\right)$$

The transfer function may be either physically or empirically derived, may be non-linear, many-to-one or probabilistic. Typical examples include electricity demand models (Thornton et al. 2016; Taylor and Buizza 2003), wind power production models (Dunning et al. 2015; Cannon et al. 2015) and damage models (McColl et al. 2012).

A more complex form of impact occurs when the simultaneous influ-ence of meteorology on several different components of an energy system becomes an important part of the impact. In this case, a transfer function exists for each component, but the impact is perceived through a combination of those components (often referred to as a 'compound impact'). An example is the residual power load of national power systems (i.e., demand net renewables), which depends on demand, solar and wind, each of which has a different sensitivity to weather.

In the simplest case, the system can be considered as a set of non-interacting energy system components, $\{E\}$, and may be written:

$$S = L\big(\{E\}\big)$$

where L is a mapping of the set of energy system components $\{E\}$ to a particular system-wide property of interest, S.

An example[1] of this is the 'merit order' model[2] of UK wholesale power price, exploring the extent to which month-ahead forecasts could be beneficial to energy-trading and risk management (Fig. 11.4a; see Lynch et al. 2014; Lynch 2016). In this example, daily ensemble forecasts of UK wind power and national total power demand are created from the European Centre for Medium-Range Weather Forecasts (ECMWF) system for several weeks in advance, and the 'residual demand' calculated.[3] The residual demand is assumed to be met by a mixture of coal and gas generation, preferentially utilising the cheapest marginal cost generators first (i.e., those bidding to produce power at the lowest price), with the wholesale power price being determined by the most expensive generation unit required to operate (Fig. 11.4b). The use of sub-seasonal weather forecasts three to four weeks ahead was shown to offer an improvement over standard industry practice for some—though not all—trading applications. This work therefore emphasised both the potential benefits of longer-range meteorological forecasts for energy, but also the need for careful evaluation of the forecast's performance *in the context for which it is being used.*

In both 'point impact' and 'compound impact' problems, the meteorological state is assumed to map directly to that of the impacted system (via a transfer function or set of transfer functions) and, although the mapping may be complicated, it is only dependent on the current meteorological state. This assumption does not hold, however, in many energy system planning and operations problems (e.g., 'optimal power flow' or 'unit commitment'). In problems of this type, there are potentially complex connections in time and space between different energy system components and to forecast the state of the impacted system accurately requires

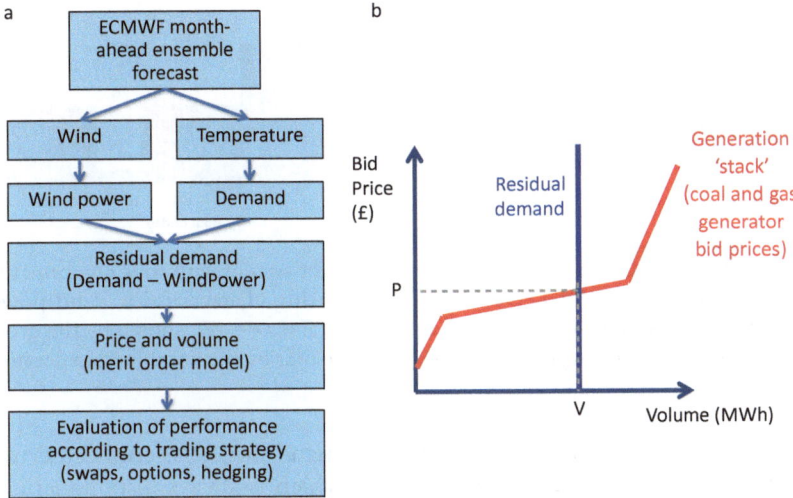

Fig. 11.4 An illustration of energy price forecasting using meteorological inputs following Lynch (2016) and Lynch et al. (2014). (**a**) A flow chart illustrating the process through which the forecast is made and evaluated. (**b**) A schematic of the 'merit order model'. In (**b**), the red curve indicates the relationship between supply and price (more expensive power stations are willing to produce as price rises, hence a positive relationship between volume and price). The blue curve indicates the relationship between demand and price (the demand for power decreases with price, but here is assumed to be perfectly price-insensitive). The intersection of the two curves sets the wholesale price and volume of power produced by the market. The qualitative shape of the supply curve produced by the two-generation type model (as fitted by Lynch (2016) to observed price data using an Ensemble Kalman filter) is indicated in (**b**). Lynch (2016) went on to demonstrate that the ECMWF-forecast based process outlined in (**a**) was able to significantly outperform equivalent forecasts using purely historical weather observations for each of wind power, demand and price (evaluated over the period December 2010—February 2014, at a 99% statistical confidence level). ECMWF stands for European Centre for Medium-Range Weather Forecasts

knowledge of both the power system's initial state and the meteorological evolution between the forecast's initialisation and its target lead time.

It is beyond the scope of the present text to discuss these problems in detail but the key concepts of a 'complex impact' on the power system can be illustrated through a conceptual model,[4] as shown in Fig. 11.5.

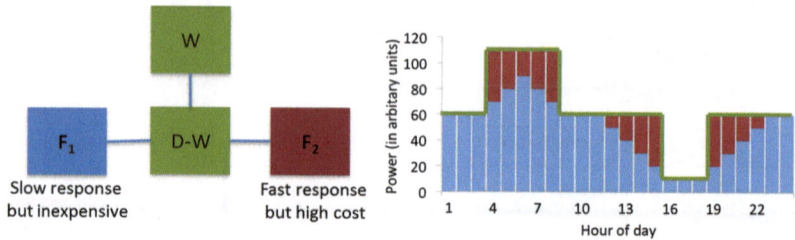

Fig. 11.5 A conceptual model of a simple power system with four components: two fossil fuel generators (F_1 and F_2) with differing characteristics, wind power generation W and demand D. Residual demand ($E = D - W$), shown by the green line on the time series (right-hand plot), must be met by the combined generation from F_1 and F_2. See main text for discussion

Consider a power system with a single demand node, D, connected to a wind power source, W, the output from which is always instantaneously utilised. As there is no storage of power, the residual demand E (i.e., demand minus wind power, $E = D - W$) must be met at all times by two fossil fuel generators, F_1 and F_2. F_1 has low fuel costs (i.e., it is cheap to generate power with F_1) but changes in its output must occur slowly, whereas F_2 has high fuel costs (i.e., it is expensive to use) but its output can change rapidly if required.

Consider further a time series of residual demand as shown by the green line on the right hand panel of Fig. 11.5. Initially, the residual demand can be met entirely by F_1—the low cost generator—but on hour 4 the residual demand rapidly increases faster than F_1 can respond and F_2 must be used to meet the short fall. Crucially, although the residual demand at hour 4 could have been determined using a transfer function applied to the instantaneous meteorological state,[5] the division of the generation used to meet this residual demand between F_1 and F_2 in hour 4 *could not* have been estimated without also knowing the *prior and future* meteorological and power system status. In this example, if the residual demand had been higher in hour 3—and hence $F_1(t = 3)$ was also higher—then more of the residual demand in hour 4 could have been met with the cheaper F_1 rather than the more expensive F_2. In effect, it is not possible to determine the value of F_1 and F_2 at a particular point in time (e.g., $t = 4$) independently of determining F_1 and F_2 over many surrounding time steps.

Thus, if one wishes to model the status of the power system at any instant, it is therefore important to correctly represent both the meteorological time series trajectory *and* the power system's time-evolving response to it. *It is not sufficient to simply apply a transfer function to an instantaneous 'snapshot' of weather in isolation from the rest of the time series trajectory to produce a full estimate of the power system's status.* In practice, the time-dependences introduced by power system response constraints are also further complicated by spatial connections introduced by transmission limitations (i.e., finite rates of power transfer between locations). In contrast to 'point forecasting', however, there has been relatively little assessment of 'time-trajectory forecasting' (or spatial patterns of co-dependent meteorological surface variables) in the meteorological research literature in either a weather-forecasting or climate modelling context. Similarly, there has also been relatively little attention paid to the quality of meteorological data used in sophisticated energy system planning and operations studies. New research is, however, beginning to tackle some of these concerns, for example, Bloomfield et al. (2016) highlight that significant errors may arise if insufficiently long weather records are used for power system planning and Pfenninger and Keirstead (2015) have provided a recent example of complex unit commitment modelling in a climate-change context.

Despite the differing levels of energy system impact complexity, many challenges in energy meteorology have similarities to other meteorological applications (e.g., insurance, water and agriculture). The need to calibrate and downscale meteorological variables from coarse prediction datasets to specific localised properties is a particularly ubiquitous problem. Direct meteorological observations of the site (for 'statistical downscaling') and 'dynamical downscaling' (with finer resolution numerical models) can assist in many circumstances, but it is especially challenging when the response of the impacted system depends on more than one meteorological input (in such cases, the co-variability of the meteorological properties may be important as well as the individual meteorological properties themselves). It is also noted that downscaling and calibration only improve the forecast if the large-scale dynamics of the system are well-simulated and, in practice, errors associated with meteorological downscaling and transfer functions are often difficult to separate (see, e.g., Cannon et al. 2017).

SUMMARY

To summarise, the transfer from 'meteorology' to 'energy' is, in many cases, highly non-linear. This has profound implications for simulation and prediction of energy system impacts, suggesting that forecast skill may be strongly influenced by the transformation from meteorological variables to energy variables. In an ideal world, this may act to either increase or decrease the skill of the forecast, depending on the specific characteristics of the forecast problem but, in practice, the skill of an *energy* forecast will often tend to be lower than the *meteorological* forecast from which it originates as errors in the transformation process will tend to compound errors in the original meteorology. Careful diagnosis is needed to identify which aspects of the forecasting system—from the meteorological prediction to its downscaling and transformation into an energy property, and finally its conversion into an end-user decision—lead to the dominant sources of error, and to focus analytical resources on the scales and processes where skill is achievable.

NOTES

1. Other examples of similar 'compound impact' problems can be found in peak-load estimation (e.g., Thornton et al. 2017) and simple models for system planning applications (e.g., load duration curves for the estimation of the optimal generation-type mix: Green and Vasilakos 2010; Bloomfield et al. 2016).
2. See, e.g., Staffell and Green (2016) for an introduction to 'merit order' concepts.
3. The residual demand is presented here as total demand minus wind power generation for simplicity. In practice, Lynch (2016) made several additional calculations, removing inflexible generators (such as nuclear) and other varying contributions (such as embedded solar and interconnectors) from the total demand.
4. See, e.g., Wood et al. (2014) and Staffell and Green (2016) for an introduction.
5. That is, a function of the form $E(t = 4) = f(\{m(t = 4)\})$.

REFERENCES

Bloomfield, H. C., Brayshaw, D. J., Shaffrey, L. C., Coker, P. J., & Thornton, H. E. (2016). Quantifying the increasing sensitivity of power systems to climate variability. *Environmental Research Letters, 11*, 124025.

Brayshaw, D. J., Troccoli, A., Fordham, R., & Methven, J. (2011). The impact of large scale atmospheric circulation patterns on wind power generation and its potential predictability: A case study over the UK. *Renewable Energy, 36*, 2087–2096.

Cannon, D. J., Brayshaw, D. J., Methven, J., Coker, P. J., & Lenaghan, D. (2015). Using reanalysis data to quantify extreme wind power generation statistics: A 33 year case study in Great Britain. *Renewable Energy, 75*, 767–778.

Cannon, D. J., Brayshaw, D. J., Methven, J., & Drew, D. (2017). Determining the bounds of skillful forecast range for probabilistic prediction of system-wide wind power generation. *Meteorologische Zeitschrift, 26*, 239–252.

Dunning, C. M., Turner, A. G., & Brayshaw, D. J. (2015). The impact of monsoon intraseasonal variability on renewable power generation in India. *Environmental Research Letters, 10*, 064002.

Green, R., & Vasilakos, N. (2010). Market behavior with large amounts of intermittent generation. *Energy Policy, 38*, 3211–3220.

Lynch, K. J. (2016). *Subseasonal forecasting for the energy sector*. PhD Thesis, University of Reading.

Lynch, K. J., Brayshaw, D. J., & Charlton-Perez, A. (2014). Verification of European subseasonal wind speed forecasts. *Monthly Weather Review, 142*, 2978–2990.

McColl, L., Palin, E. J., Thornton, H. E., Sexton, D. M. H., Betts, R., & Mylne, K. (2012). Assessing the potential impact of climate change on the UK's electricity network. *Climatic Change, 115*, 821–835.

Pfenninger, S., & Keirstead, J. (2015). Renewables, nuclear or fossil fuels? Comparing scenarios for the Great Britain electricity system. *Applied Energy, 152*, 83–93.

Staffell, I., & Green, R. (2016). Is there still merit in the merit order stack? The impact of dynamic constraints on optimal plant mix. *IEEE Transactions on Power Systems, 31*, 43–53.

Taylor, J. W., & Buizza, R. (2003). Using weather ensemble predictions in electricity demand forecasting. *International Journal of Forecasting, 18*, 57–70.

Thornton, H. E., Hoskins, B. J., & Scaife, A. A. (2016). The role of temperature in the variability and extremes of electricity and gas demand in Great Britain. *Environmental Research Letters, 11*, 114015.

Thornton, H. E., Scaife, A. A., Hoskins, B. J., & Brayshaw, D. J. (2017). The relationship between wind power, electricity demand and winter weather patterns in Great Britain. *Environmental Research Letters, 12*, 064017.

Wood, A. J., Wollenberg, B. F., & Sheble, G. B. (2014). *Power generation, operation and control*. Hoboken, NJ: Wiley. 632pp.

Open Access This chapter is distributed under the terms of the Creative Commons Attribution 4.0 International License (http://creativecommons.org/licenses/by/4.0/), which permits use, duplication, adaptation, distribution and reproduction in any medium or format, as long as you give appropriate credit to the original author(s) and the source, a link is provided to the Creative Commons license and any changes made are indicated.

The images or other third party material in this chapter are included in the work's Creative Commons license, unless indicated otherwise in the credit line; if such material is not included in the work's Creative Commons license and the respective action is not permitted by statutory regulation, users will need to obtain permission from the license holder to duplicate, adapt or reproduce the material.

CHAPTER 12

Probabilistic Forecasts for Energy: Weeks to a Century or More

John A. Dutton, Richard P. James, and Jeremy D. Ross

Abstract Quality of service and fiscal success in the energy industry often depend on how well meteorological information and forecasts are used to manage risk and opportunity. On the subseasonal to seasonal (S2S) timescales, a disciplined strategy allows decision makers to counteract predicted adverse climate variations in the coming weeks or months with action or financial hedges. Calibrated S2S probabilistic forecasts from some providers have sufficient skill that they engender confidence in the statistical consequences of acting. On the scale of several or more decades ahead, probabilistic outlooks can guide strategic planning and capital expenditures in directions that will ensure long-term resilience to climate change. In both cases, the probabilities are generated by statistical analysis of ensembles of supercomputer forecasts or climate change scenarios.

Keywords Probabilistic forecasts • Subseasonal and seasonal climate prediction • Climate change • Resilience • Energy industry

J.A. Dutton (✉) • R.P. James • J.D. Ross
Prescient Weather Ltd, State College, PA, USA

© The Author(s) 2018
A. Troccoli (ed.), *Weather & Climate Services for the Energy Industry*,
https://doi.org/10.1007/978-3-319-68418-5_12

161

INTRODUCTION

The energy industry has a voracious appetite for meteorological information on many time and space scales, and both the quality of service and fiscal performance depend on how well the information is used to manage risk and take advantage of opportunity. Today the various components of the industry can combine probabilistic information with sophisticated decision methods to produce predictable and desirable statistical results (Dutton et al. 2013, 2014).

On the subseasonal to seasonal (S2S) timescale, a disciplined strategy allows decision makers to counteract likely adverse events in the coming weeks or seasons with action or financial hedges. On the scale of several or more decades ahead, probabilistic outlooks can guide strategic planning and capital expenditure in directions that will ensure long-term resilience to climate change.

The most useful S2S forecasts and climate change outlooks are probability distributions created from evolving ensembles of forecasts generated by supercomputers calculating tens of forecasts simultaneously by perturbing initial conditions, model characteristics, or boundary conditions. The predicted probability distributions allow decision makers to distinguish between likely and unlikely conditions or events and to respond appropriately.

Indeed, the predicted probability distributions are analogues of the frequency distributions that are used to describe the climatological averages and volatility of energy-critical variables such as temperature. As illustrated by Fig. 12.1, the predicted distributions can show that significant departure from climatological conditions is expected and that action may be warranted.

SUBSEASONAL AND SEASONAL CLIMATE PREDICTION

S2S forecasts[1] covering weeks to three or six months do not themselves produce benefits in the energy industry or in other activities. Making them useful requires a process to convert forecasts into actionable information and to estimate the consequences of acting on the forecasts. A National Research Council report (NRC 2016) offers a research agenda to improve S2S forecasts.

The energy industry seeks forecasts of future events on S2S timescales in order to minimize adverse results or take advantage of opportunity. For

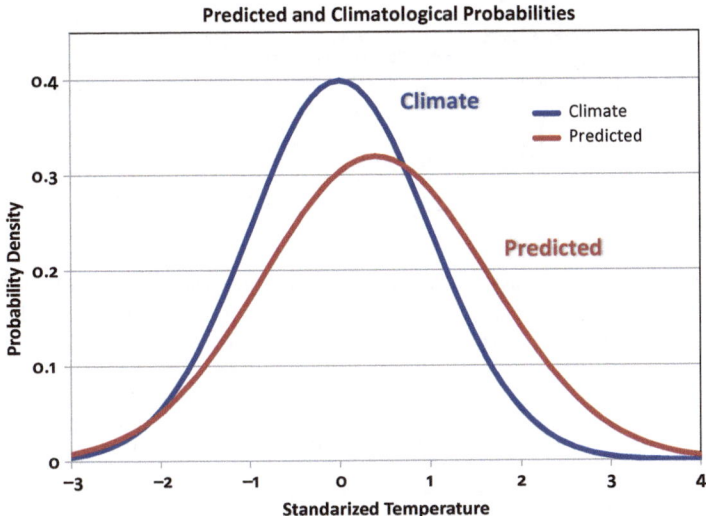

Fig. 12.1 Comparison of predicted and climatological standardized temperatures for a S2S forecast. The area between the climate and predicted densities represents the probability of temperatures warmer than those expected from climate, which would be adverse for winter for an electric utility

example, electric utilities usually consider warm winter temperatures and cool summer temperatures as adverse because income may not meet expectations and thus they may attempt to ensure financial stability with hedges. The possible future states considered in S2S temperature and other forecasts are often divided into three categories: below normal, nearly normal, and above normal, each of frequency one-third in the historical record for each location and time period. S2S forecasts usually provide a predicted probability for each of the three terciles.

The key question users often ask is: *At what predicted probability should I act?* The better question is: *What consequences can I expect if I act at a predicted probability equal to p?* The critical resource for bridging the gap between forecasts and decisions to act is a reliable description of the performance of the forecast system. Then it becomes possible to link statistical summaries of the consequences to various values of predicted probabilities and to answer the question about consequences of action. Figure 12.2 provides a description of such a forecast system.

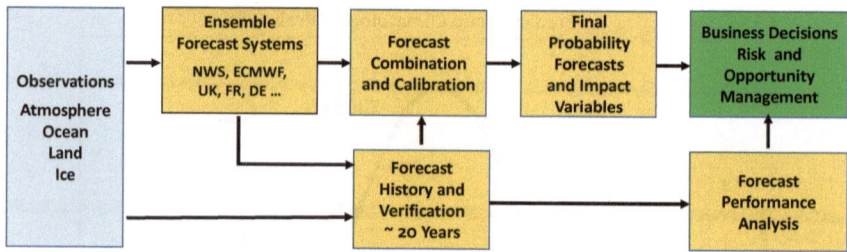

Fig. 12.2 A S2S forecast system that uses the forecast history and verification data to optimize new forecasts in a calibration and combination process. The same data leads to forecast performance statistics that inform the business decisions

Table 12.1 The business model for computing the consequences of forecasts and hedges

		Events	
		Adverse	Favorable
Fore-casts	Adverse	$R_a + H - C(H)$	$R_f - C(H)$
	Favorable	R_a	R_f

Separating the range of predicted variables into two classes—adverse and favourable—simplifies and illuminates the interaction of forecast performance and business decisions. Let us consider returns R_f for favourable conditions and R_a for adverse with a loss L being the difference. We consider a hedge for predicted adverse conditions that pays H if they prevail and costs $C(H)$ to establish. Then the contingency table that describes the four possibilities is shown in Table 12.1.

Now we turn to the forecast performance statistics to compute the probabilities of occurrence of each possibility when the adverse case is predicted with probability equal to or greater than p.

We divide the range $[0, 1]$ of predicted probabilities into 10 bins with centres at $0.05, 0.15, \ldots 0.95$ and from the history of forecasts and verification we count for each bin the numbers $V(p)$ and $X(p)$ of correct and incorrect adverse forecasts with a forecast for adverse considered correct if the subsequent observed verification value is in the adverse range. The

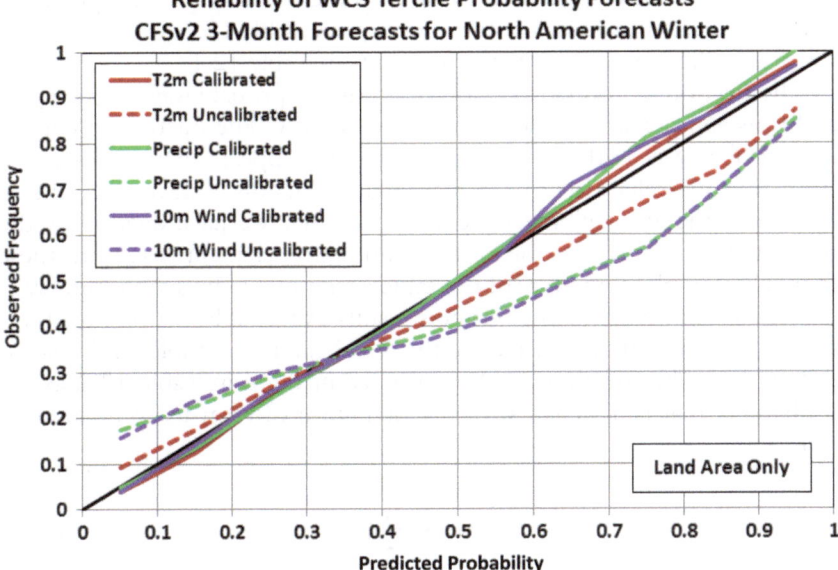

Fig. 12.3 A reliability diagram for WCS forecasts of temperature, precipitation, and wind speed for the North American winter, illustrating the improvement in reliability achieved by calibration. The data for above and below normal have been combined to create a single curve for each variable and thereby simplify the diagram

total number of forecasts $S(p) = V(p) + X(p)$ in each bin is known as the sharpness of the forecasts. We divide these three quantities by the total number N of forecasts and then have the corresponding ratios $v(p)$, $x(p)$, and $s(p)$. The reliability is defined to be $r(p) = v(p)/s(p)$ and thus is a ratio of the number of correct forecasts to the total number of forecasts and is an important indicator of forecast performance.[2] Figure 12.3 shows a reliability diagram for a set of World Climate Service (WCS) forecasts.[3]

To describe expected outcomes for action at all predicted probabilities $p_p \geq p$, we sum over this range and so the fraction $f_a(p)$ of adverse forecasts and the fraction $F_a(p)$ of correct forecasts are:

$$f_a(p) = \sum_{y \geq p} s(y), \ F_a(p) = \left(\sum_{y \geq p} v(y) \right) / f(p) \quad (12.1)$$

and we use these two quantities along with the climatological frequency n_a of adverse events to complete the contingency Table 12.2, which contains the probabilities associated with the events in Table 12.1.

The fraction of adverse forecasts at predicted probabilities $p_p \geq p$ is $f(p)$ and thus appears as the Sum of the Adverse row. The fraction of correct adverse forecasts is the product $f(p)F(p)$ and appears in the Adverse × Adverse matrix element. The climatological fraction of adverse events n_a appears as the Sum of Adverse events and since we have divided by the total number of forecasts the Sum × Sum matrix element is 1. With these four values in place, the rest of the table is completed by simple algebra.

Now we can describe the business results expected by acting on a forecast of adverse conditions. Define the 2 × 2 matrix in Table 12.1 shaded yellow as the business model M and its companion in Table 12.2 as the probability matrix P. Then with the definition of term-by-term summation as

$$A \circ B = \sum_{i=1}^{2} \sum_{j=1}^{2} A_{i,j} B_{i,j} \qquad (12.2)$$

we can compute the expected revenue $R(p)$ and its variance $V(p)$ obtained when acting on $p_p \geq p$ as

$$R = P \circ M, \qquad V = P \circ M^2 - R^2 \qquad (12.3)$$

in which the elements of M^2 are the squared elements of M. Here R and V are functions of the variables in Table 12.1 and of the predicted probability p via the functions in Table 12.2.

To obtain quantitative estimates, we must have suitable representations of the forecast performance functions $f_a(p)$ and $F_a(p)$, as illustrated in Fig. 12.4 for WCS forecasts of temperature, precipitation, and wind for the North American winter. The computations of expected return and variance are simplified by modelling the summands in (12.1) with beta functions, converting the sums in (12.1) to integrals and performing the integration, and thereby obtaining analytical expressions for $f_a(p)$ and $F_a(p)$.

To complete the analysis, we need estimates of the cost of various hedges. For hedges that pay when the observed verification value falls

Table 12.2 The frequencies of events in the forecast contingency table with the dependence on p of the skill functions omitted for brevity

		Events		
		Adverse	Favorable	Sum
Fore-casts	Adverse	$f_a F_a$	$f_a(1-F_a)$	f_a
	Favorable	$n_a - f_a F_a$ n	$1 - n_a - f_a(1-F_a)$	$1 - f_a$
	Sum	n_a	$1 - n_a$	1

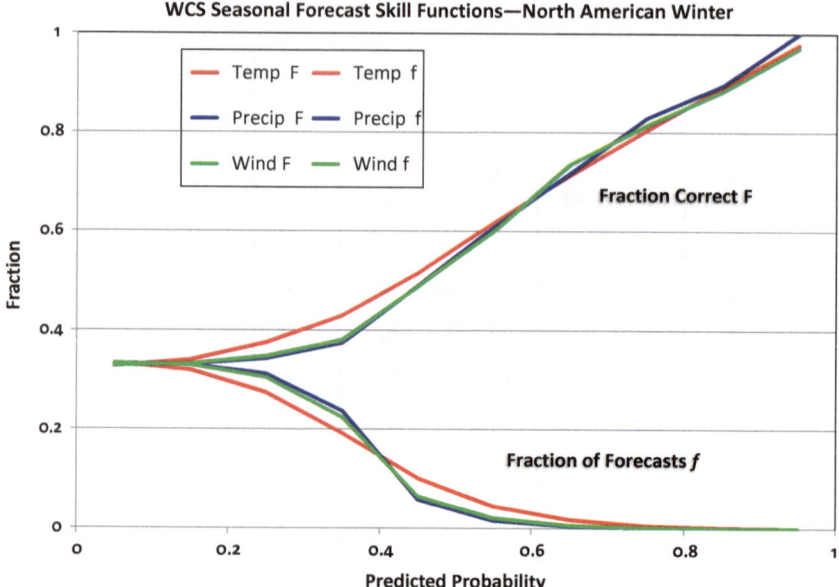

Fig. 12.4 Forecast performance functions for WCS forecasts for temperature, precipitation, and wind for the North American winter

anywhere in the adverse tercile, some sellers use $C(H) = \bar{H} + \eta\sigma_H$ with $\bar{H} = n_a H$ and $\sigma_H^2 = n_a H^2 - \bar{H}^2 = n_a(1 - n_a)H^2$ and often select $\eta = 1/4$.

For example, the WCS maintains and displays detailed information about the skill and reliability of its S2S forecasts relative to the terciles (Dutton et al. 2013; James et al. 2014). Now the WCS is combining forecast performance records with a model of the hedging process to create a hedge advisor, shown in Fig. 12.5, that provides expected returns and volatilities for hedges put in place at various predicted probabilities of adverse conditions (Dutton et al. 2015). The return $R(H, p,...)$ and the volatility $\sqrt{V(H, p,...)}$ (standard deviation) are plotted parametrically as functions of H for various values of predicted probability p for warm North American winters with $R_f = 100$ (units arbitrary) and $L = 33$.

These plots thus take explicit account of the historical skill of the forecasts, and thus both buyers and sellers of hedges can act with some confidence about results expected over a number of cases.

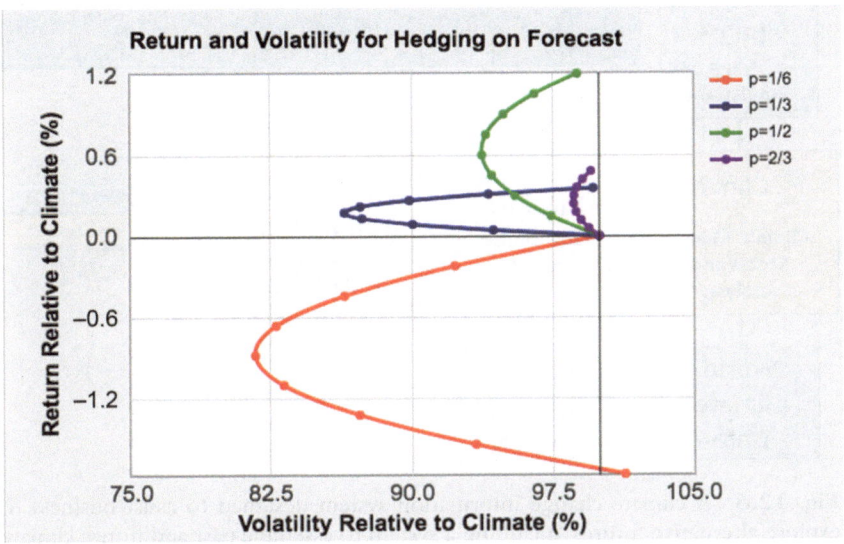

Fig. 12.5 The WCS hedge advisor compares volatility and return for several predicted probabilities of adverse events to those expected from the climatological frequency of the same event. The dots indicate hedges at increments of one-quarter of the loss in adverse conditions. The minimum volatility occurs for hedges close to that loss. In this illustration, the skill of the forecasts puts the seller of hedges using $\eta = 1/4$ at a financial disadvantage for predicted probabilities of $1/3$ or greater

As another example, Vitart (2014) provides a summary of the skill of the S2S forecasts of the European Centre for Medium-Range Weather Forecasts (ECMWF). The WCS combines and calibrates these ECMWF forecasts with the S2S forecasts of the US National Weather Service (NWS) to form the multi-model forecasts and performance statistics discussed above. The calibration compares some three decades of retrospective forecasts for previous years with the corresponding verification to develop statistical methods for improving the current forecasts.

Climate Change Probabilities

Simulations of twenty-first century climate change on the scale of decades or more in the future provide users with an entirely different challenge related to long-term business strategy and capital investment. On this

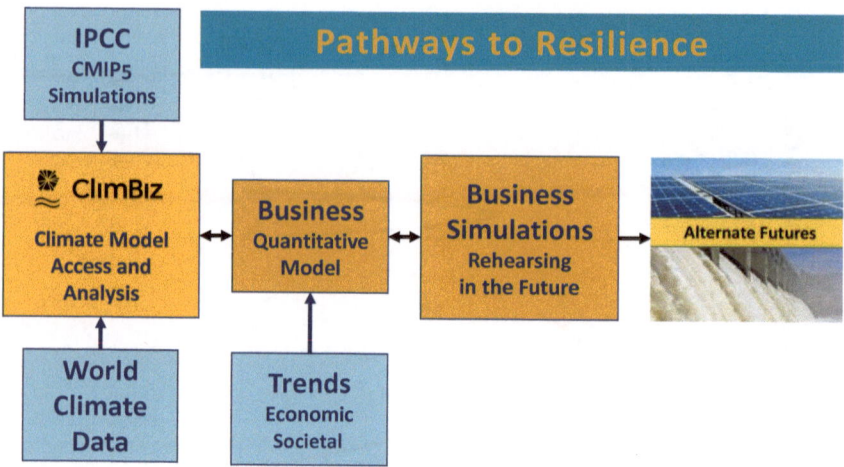

Fig. 12.6 A climate change information system designed to assist business to explore alternative futures, including a system to assemble past and future climate information, a quantitative model of the business, and a system for computing business simulations

scale, uncertainty prevails in all aspects of the energy industry, including environmental variables involved in creating demand and generating power, in technological advance, in prices of fuel or equipment, in the changing numbers and needs of the customers, and in evolving regulation in response to awareness of climate change.

A comprehensive strategy for addressing this challenge is illustrated in Fig. 12.6 which shows how data about observed and future climate can be combined with a quantitative business model to generate simulations of future performance.[4] The strategy has three components: a source of information about past and future climates, a quantitative business model, and a set of business simulations.

For future climate information, we presently use the climate simulations of 16 national and international modelling centres prepared for the Climate Model Intercomparison Project 5 (CMIP5) (Taylor et al. 2012) for the fifth report of the Intergovernmental Panel on Climate Change (IPCC 2013). Using these simulations, we can create probability distributions for environmental variables that depict climate evolution and variation as forced by greenhouse gas emission scenarios designed to cover a wide range of possibilities.

With such climate change simulations, there are no forecast verifications and only performance statistics for versions run for the twentieth century with quite different forcing. Nevertheless, the producers and consumers of energy can use probabilistic information from the climate change scenarios to examine the relevance and resilience of their business models and strategies. They can prepare now for change that, however unclear, is certain to come.

A mathematical and numerical model of a hypothetical utility, the Virtual Electric Power Company (VEPCO),[5] illustrates how the strategy of Fig. 12.6 might be implemented. An influence diagram in Fig. 12.7, constructed following Brown (2015), describes a business model that is combined with the evolving probability distributions of temperature, insolation, and precipitation for moderately severe climate change obtained from CMIP5 climate simulations to estimate demand and the availability of solar and hydro power. VEPCO plans an increased reliance on solar and hydro power because of decreasing costs expected for these renewables,

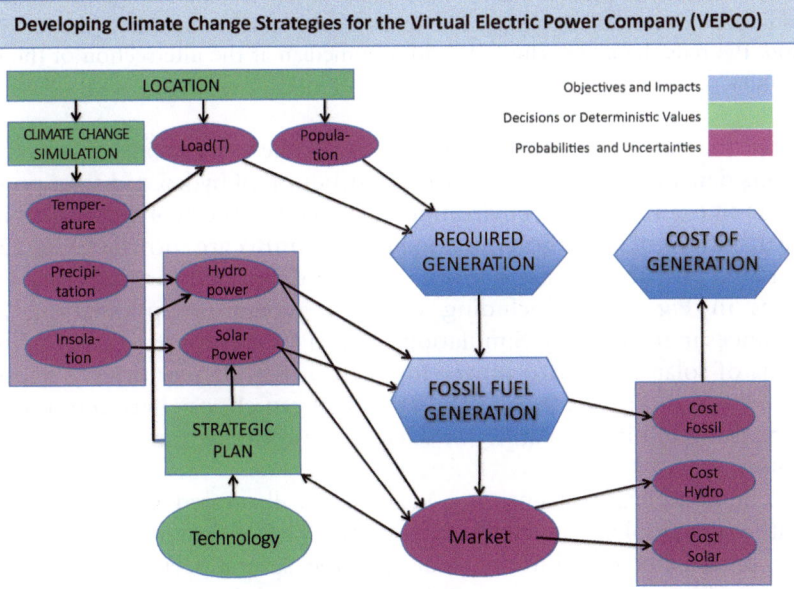

Fig. 12.7 The business model, constructed as an influence diagram, used to generate climate change scenarios for the hypothetical utility VEPCO

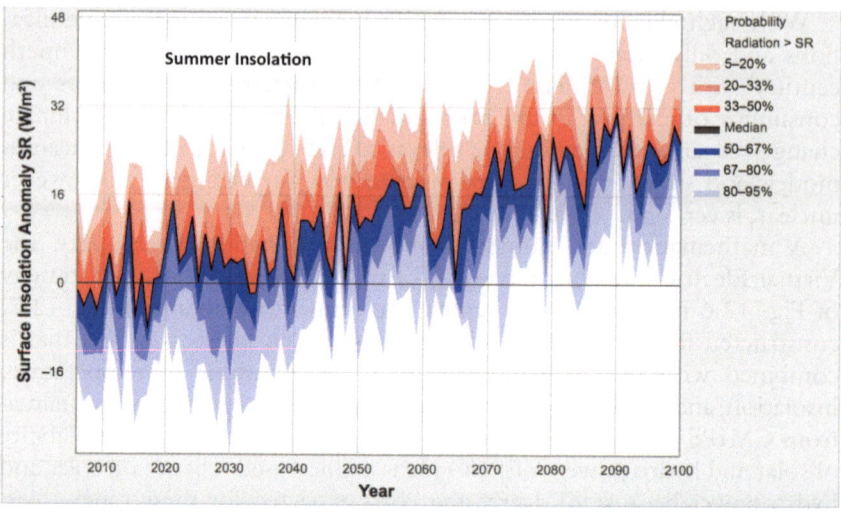

Fig. 12.8 Evolving probability of summer insolation for VEPCO, from one of the IPCC climate simulations for moderately severe climate change (6 watts/m² additional greenhouse heating). The blues are the small-value side of the distribution, the reds the large-value side with the median at the intersection of the two colours

while the cost of fossil power increases. In the simulations, fossil power meets demand remaining after the contribution of hydro and solar power.

As an example of the environmental variables, the evolving probability distributions of insolation obtained from ClimBiz are shown in Fig. 12.8. Suitable probability distributions must be developed for the other variables in Fig. 12.7, including costs of generation, population, and advances in technology. Simulations of VEPCO response to 15 combinations of solar and hydro power for each of the 20-year double decades are shown in Fig. 12.9. And thus VEPCO can choose between minimal expense or minimal volatility or select some combination it expects to be optimum.

The complexity of ensuring resilience is illustrated by this example. Rather than looking at simple statistics, the VEPCO planners can combine the probability distributions that describe several scenarios of climate change from mild-to-severe with distributions describing the potential range of customer needs, technology, policy imperatives, and market

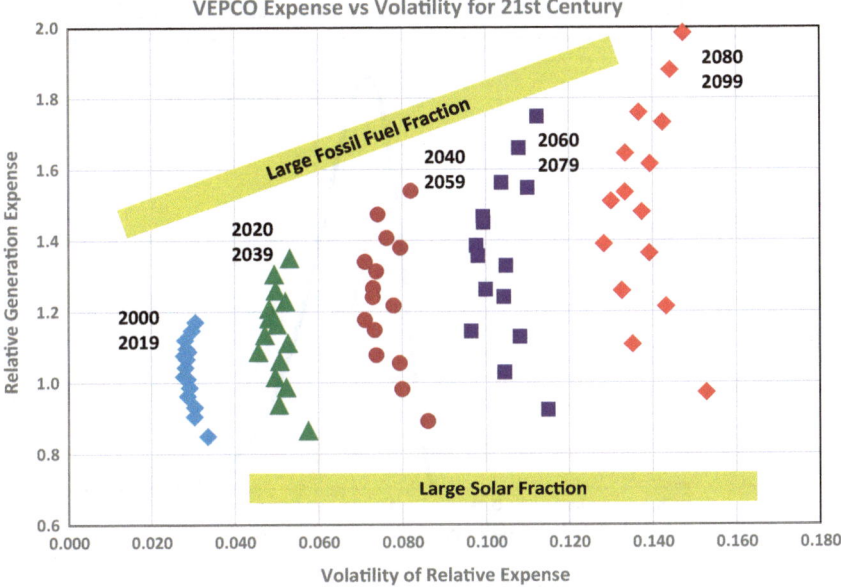

Fig. 12.9 Expense versus volatility relative to present conditions for 15 selections of solar and hydro fractions of generation for five double decades in the twenty-first century, giving VEPCO a range of possibilities for minimizing expense or minimizing volatility. The relative volatility on the x-axis is the standard deviation of the relative expense on the y-axis

forces as the twenty-first century evolves. Thus, they must combine a comprehensive model of the business and probabilistic models of a variety of forces that may drive change in the business. Sampling from all of these probability distributions will produce an immense amount of data. But all the individual scenarios will combine into smooth probability distributions that depict both likely events in the centre of the distribution and the likelihood in the tails of both adverse and favourable events for which VEPCO must be alert and be prepared to act if necessary.

Being ready for whatever comes is the key benefit of resilience and of examining possible future events through the window of probabilities that describe both their likelihood and uncertainty. As summarized by Hamel and Välikangas (2003): "In a truly resilient organization, there is plenty of excitement … but no trauma."

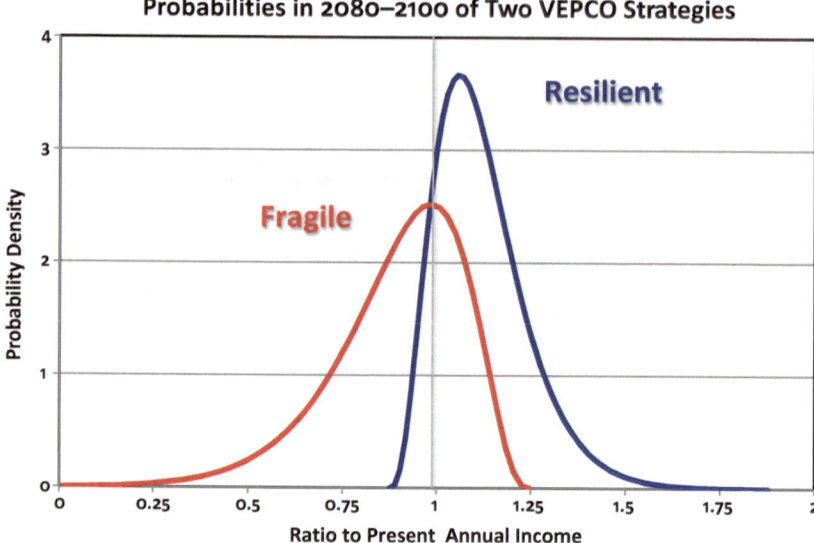

Fig. 12.10 Probability densities associated with VEPCO strategies that would imply favourable or unfavourable prospects for the end of the century. The performance is measured by a ratio of 2080–2100 income to present-day income, both in present-day values. Fragile is used in the sense of vulnerability to volatility (Taleb 2012), resilience for anti-fragile. The resilient density is relatively thin and favours positive income and therefore is robust

Hypothetical probabilities for two different strategies for VEPCO are illustrated in Fig. 12.10, one resilient and one fragile and undesirable. With a business simulation driven by scenarios of climate change and the evolution of economic forces, VEPCO can create probabilistic portraits of its operating variables for the decades to come and identify decisions and action that will ensure resilience. It will be "rehearsing the future", as advocated by Schwartz (1996).

And rehearsing successfully to ensure resilience in the decades ahead will confer the ultimate competitive advantage in the energy and other industries.

CONCLUSION

This examination of probability forecasts for energy on timescales ranging from weeks to a century or more has demonstrated that decisions must be two-dimensional and consider both a measure of return or cost and a measure of risk or volatility related to the variance of return or cost, thus echoing the Nobel Prize-winning conclusions of Markowitz (1952) about investment selection.

Hedging adverse forecasts on the S2S scale leads to a range of choices, with maximum return accompanied by maximum volatility compared to minimum volatility with reduced return. Assessment of climate change strategies focused on generation cost produced a similar set of choices for a range of configurations and capital investment commitments for a virtual utility. Finding the pathways to resilience across a variety of potential climate change trajectories thus requires examining a collection of scenarios and then comparing overall return or cost to overall volatility across the full range of potential variation.

Achieving resilience on any timescale has three critical components: forecasts or scenarios for a future period, a model of the business that will yield results as a function of possible hedges or other decisions and actions, and a history or other means of assessing the quality of the forecasts. For S2S forecasts, we created a generic business model and showed how forecast skill functions then produced analytical and numerical comparisons of return and volatility for various predicted probabilities and hedges. For climate variability, the assessment of the expected accuracy and relative value of various climate simulations remains a signal challenge for the climate research community. Until that challenge is met, it seems that the best statistical strategy is to use as many simulations as possible and scale them to a common climatological base over a decade or two. This may produce overly broad probability distributions, but that is preferable to having them too narrow and producing overconfident estimates.

In summary, energy firms can control their statistical future for S2S timescales if sufficiently skilful forecasts are available; for climate change, they can explore a range of statistical futures in search of the pathways to resilience.

Acknowledgements This chapter is based on Prescient Weather Ltd research supported by the US National Oceanic and Atmospheric Agency with Contracts WC133R-11-CN-0147 and WC-133R-16-CN-0103 and by the US Department of Energy with award DE-SC0011284.

Notes

1. S2S forecasts are currently available from the US National Weather Service (NWS), Environment Canada, the European Centre for Medium-Range Weather Forecasts (ECMWF) and the commercial World Climate Service which combines the NWS and ECMWF forecasts into a multi-model ensemble. The European Copernicus project is offering S2S forecasts from a number of national forecast centres and the NWS is coordinating development of an experimental subseasonal component of the North American Multi-Model Ensemble (NMME). There may be others of which the authors are unaware.
2. We would consider a forecast for rain reliable if it rains on one-third of the days for which we predicted a probability of one-third for rain.
3. A collaborative effort of Prescient Weather in the US and MeteoGroup, a global weather information firm with headquarters in London.
4. This is part of the development by Prescient Weather of a Climate Change Information System for Business and Industry (ClimBiz) sponsored by the US Department of Energy.
5. The virtual VEPCO shares its acronym with the real but unrelated Virginia Electric and Power Company.

References

Brown, R. D. (2015). *Business case analysis with R*. Retrieved from https://leanpub.com/bizanalysiswithr

Dutton, J. A., James, R. P., & Ross, J. D. (2013). Calibration and combination of dynamical seasonal forecasts to enhance the value of predicted probabilities for managing risk. *Climate Dynamics, 40*, 3089–3105.

Dutton, J. A., James, R. P., & Ross, J. D. (2014). A probabilistic view of weather, climate, and the energy industry. In A. Troccoli, et al. (Eds.), *Weather matters for energy* (pp. 353–378). New York: Springer.

Dutton, J. A., James, R. P., & Ross, J. D. (2015). Bridging the gap between sub-seasonal and seasonal forecasts and decisions to act. *AMS Annual Meeting*. Phoenix, AZ. Retrieved January 7, 2015, from https://ams.confe12.com/ams/95Annual/webprogram/Paper260171.html

Hamel, G., & Välikangas, L. (2003). The quest for resilience. *Harvard Business Review, 81*, 52–63.

James, R. P., Ross, J. D., & Dutton, J. A. (2014). Skill of a new two- to -six week forecast system. *AMS Annual Meeting*. Atlanta. Retrieved February 4, 2014, from https://ams.confe12.com/ams/94Annual/webprogram/Paper233913. html

Markowitz, H. (1952). Portfolio selection. *Journal of Finance, 7*, 77–91.

National Research Council. (2016). *Next generation earth system prediction: strategies for subseasonal to seasonal forecasts*. Washington, DC: National Academies Press. 336 pp.

Schwartz, P. (1996). *The art of the long view*. Reprint with new user's guide. New York: Crown Business. 272 pp.

Taleb, N. N. (2012). *Antifragile: Things that gain from disorder*. New York: Random House. 519 pp.

Taylor, K. E., Stouffer, R. J., & Meehl, G. A. (2012). An overview of CMIP5 and the experiment design. *Bulletin American Meteorological Society, 93*, 485–498.

Vitart, F. (2014). Evolution of ECMWF sub-seasonal forecast skill scores. *Quarterly Journal of the Royal Meteorological Society, 140*, 1889–1899.

Open Access This chapter is distributed under the terms of the Creative Commons Attribution 4.0 International License (http://creativecommons.org/licenses/by/4.0/), which permits use, duplication, adaptation, distribution and reproduction in any medium or format, as long as you give appropriate credit to the original author(s) and the source, a link is provided to the Creative Commons license and any changes made are indicated.

The images or other third party material in this chapter are included in the work's Creative Commons license, unless indicated otherwise in the credit line; if such material is not included in the work's Creative Commons license and the respective action is not permitted by statutory regulation, users will need to obtain permission from the license holder to duplicate, adapt or reproduce the material.

CHAPTER 13

Lessons Learned Establishing a Dialogue Between the Energy Industry and the Meteorological Community and a Way Forward

Laurent Dubus, Alberto Troccoli, Sue Ellen Haupt, Mohammed Sadeck Boulahya, and Stephen Dorling

Abstract Work at the nexus between energy and meteorology aims at integrating meteorological information into operational risk management and strategic planning for the energy sector, at all timescales, from

L. Dubus (✉)
EDF – R&D, Applied Meteorology Group, Chatou, France

A. Troccoli
World Energy & Meteorology Council, c/o University of East Anglia, Norwich, UK

S.E. Haupt
NCAR and WEMC, Boulder, CO, USA

M.S. Boulahya
WEMC, Toulouse, France

S. Dorling
UEA and WEMC, Norwich, UK

© The Author(s) 2018   179
A. Troccoli (ed.), *Weather & Climate Services for the Energy Industry*,
https://doi.org/10.1007/978-3-319-68418-5_13

long-term climate change and climate variability to shorter term local weather. Weather and climate risk management can be a powerful instrument for development—not only for building energy system resilience and thus mitigating the effects of adverse events but also for ensuring that opportunities for enhanced system efficiency are exploited. The collaboration between energy and meteorology has a long history but has recently been strengthening, particularly in response to the new challenges posed by climate change and the necessary development of low-carbon energy systems. An efficient integration of high-quality weather and climate information into energy sector policy formulation, strategic planning, risk management and operational activities now, more than ever, requires improved understanding and communication between energy and meteorology specialists and decision makers.

Keywords Climate change • Energy systems • Meteorology • Risk management • Strategic planning • Resilience • Weather and climate services

Lessons Learned in Energy and Meteorology

Examples and lessons learned have been presented in the preceding chapters. Here we attempt to summarize them, highlighting the key messages in the interaction between the energy sector and the meteorology community. The objective is to strengthen this relationship so as to achieve improved resilience and efficiency of energy systems, informed by weather, water and climate services, based on a strong scientific foundation.

Improving the Communication Between Providers and Users

As advocated by the UN's Global Framework for Climate Services[1] (GFCS) (WMO 2011, 2017), evaluating the benefits of a new meteorological product for the energy industry requires good understanding between all the actors along the energy systems value chain. But, even more importantly, a fluid communication is extremely important at the early stage of the process, to understand the needs, propose the relevant solutions and work in a co-design approach. This is a field that has been further explored in depth only in the last several years, through a range of initiatives and projects, in particular but not only via the development of climate services. The EUPORIAS project,[2] for example, developed semi-operational climate services underpinned by a co-design approach. The co-design of climate services, understood as the process through which the service is defined and developed together with the end users (and other relevant actors), is perceived as an essential step for delivering successful services that adequately respond to users' needs and requirements with regard to climate data and information (Mauser et al. 2013; Troccoli et al. 2010; WMO 2011, 2017). One essential component of the co-design approach is an effective engagement and communication between the providers and the users (Brooks 2013) alongside other critical aspects such as fully understanding the scope of the climate service, the involvement of all relevant actors, a degree of flexibility and iteration in the developmental process and the continuous evaluation of the service being developed (Buontempo et al. 2014).

In the last decade or so, the links between energy and meteorology have also been developed in several international conferences, with the organization of specific energy meteorology symposia. These are now established at events such as the American Meteorological Society annual meeting[3] and the European Meteorological Society conference,[4] where the energy meteorology session has been more and more popular in the last few years. But academics still represent the largest portion of attendees at these meetings, with fewer people from the energy industry. Some more specific conferences are successful in bringing together academic and industry people (Wind Europe[5] is one of them), whereas targeted working groups (e.g. the Utility Variable Integration Group, UVIG[6]) are very successful in getting scientists, private sector service providers and energy practitioners focused on specific problems.

To our knowledge, the International Conference on Energy & Meteorology[7] (ICEM) is at present the only sustained process within the last ten years aiming at bringing meteorology and energy experts together, with the goal to cover both weather, water and climate sciences and services, and all the fields of activity in the energy value chain, even if, due to the initial structure of the network, the wind and solar aspects of the power sector were dominant at the beginning. Since its first edition in 2011, a growing network of specialists working at the nexus between energy and meteorology (weather, water, climate sciences and services) such as energy regulators, economists, planning officers, water experts, financial and insurance brokers, utility engineers, transmission and distribution operators, meteorologists, climatologists, service providers, policy makers as well as energy industry executives, have been gathering every two years. Building on its successes, ICEM's sustained process is now providing a premium international platform with excellent networking opportunities amongst the ca. 200 participants at each conference as well as a source of the state-of-the-art in the science, policy, planning and operations in energy and meteorology. Based on the discussions during ICEMs, and parallel activities in between, an international, non-profit organization was established in 2015 to go one step further: the World Energy and Meteorology Council[8] (WEMC) is devoted to promoting and enhancing the interaction between the energy industry and the weather, water, climate and broader environmental sciences community as the stakeholders of a resilient energy services value chain under an ever changing climate. Both ICEM and WEMC were presented in Chap. 5 of this book.

Improving Decision-Making Processes

Improving decision-making processes is crucial and relates to the effective integration of improved data/forecasts/products developed operationally by the meteorology community. When the target is to improve the quality and accuracy of an existing product purely from a meteorological perspective, the improvement tends to be easier to achieve. This is the case, for instance, if an improved version of a weather model reduces the forecast error for air temperature. However, with the additional step of using the meteorological variable in a specific context (e.g. the use of air temperature to compute energy demand) or in a more complex manner (e.g. using a probabilistic forecast instead of a deterministic one), the improvement is more difficult to assess. Not only is the decision process changed, because different or additional information becomes available, but also decision-making tools may need to be adapted, or even fully redefined to be able to use the new information. Decision-making tools and processes then become more complex and in most cases the underlying meteorological product development or improvement represents only a portion of the final decision or outcome.

The first implication of the interplay between the meteorological input and the final decision is that more time is required to develop the meteorological product as it must fit the decision process in the best possible way, even if both sides—the meteorological product development and the energy decision—can be addressed in parallel, at least partially.

The second implication is that, prior to the adoption of the new product, its added value must be demonstrated, and hence the methodology to enable this demonstration needs to be defined at an early stage. This requires definition of the goal to be pursued and the objective criteria that will allow the assessment of the solution.

And then the third implication is that specific people and skills are required in the process. These experts must, collectively, be able to understand every component of the energy services value chain and must be able to interact with each other in a dynamic way covering the spectrum from the meteorological side to the final energy services at the user level. Most of all, there is a need for experts with skills to implement an objective method to evaluate the benefits of the new product. Evaluating a decision-making process, and if/how/why it should or should not be changed, is a specialist field in itself. In some companies, these skills do not exist and external experts are therefore required.

In particular, it must be shown objectively that the new decision-making process, and not only the new meteorological product, adds some value to the final decision. It requires a proof that the new approach is more beneficial than the existing one. This means that the evaluation process must be completed from the application point of view and not only from the meteorological perspective. A common mistake has been for meteorologists alone to evaluate a meteorological product from the meteorological point of view and to decide on its usefulness based solely on the meteorological performance assessment, as mentioned above. This can be misleading. Indeed, even small improvements in meteorological forecasts can result in significant added value from the end user point of view, as the transformation from meteorology to energy can be nonlinear. In addition, there is generally an asymmetry in weather-dependant processes: for instance, energy demand is very sensitive to cold temperature in France in winter, but not so much to mild temperatures (Dubus 2014). A slight improvement in cold temperature event forecasts can therefore be very valuable, while a larger improvement on mild temperature events might be irrelevant in this context.

Aside from asymmetries, and more generally nonlinearities, in the meteorology-energy transfer functions, an important consideration when focusing efforts in meteorological model improvements, is that the most benefit is not always where one would commonly expect it, such as in the performance of simulating extreme events. A case in point, in the context of construction and operations and maintenance of offshore wind farms, is that enhancements to the performance of wave models should be focused most usefully upon the narrow wave height range which determines whether work can or cannot be safely conducted, rather than on very large or very small waves (Dorling and Bacon 2017). As a generic principle, the evaluation should then always be made all along the chain, from the meteorological input to the final decision, and compared to the user's current practice. The final added value will be a mix of the meteorological value and how it enhances the user decisions. This approach has also recently been implemented more frequently in government sponsored projects (Haupt et al. 2017).

It becomes even more complex when the new meteorological solution changes in nature. Many times we have heard dialogues in which meteorologists, to answer a temperature forecast quality issue for two weeks ahead, emphasized that the user should move from deterministic forecasts to probabilistic forecasts because, due to uncertainties in the initial conditions and to nonlinear effects in the equations, and many other excellent scientific arguments, it does not make sense to use deterministic forecasts beyond three to four days' lead time. End of discussion. But moving from a deterministic temperature forecast to an ensemble prediction is not a mouse-click story when one considers the whole decision process. First there are computational issues, in particular in large companies, where there are often many dependencies between different models, decision tools and processes and reporting tools. But the most significant barrier often comes from the users' reluctance to adopt new kinds of information, especially if it is probabilistic rather than deterministic. Efficient and informed use of probabilistic weather and climate forecasts has significantly advanced in the energy sector, but there is still a natural mistrust, which is quite surprising as energy people are used to dealing with uncertainty for other variables, especially in finance and market operations. Therefore, training and education about probabilistic weather and climate forecasts must remain a key component for the effective integration of better weather and climate information in energy system decision-making processes.

Looking Ahead in Energy and Meteorology

As exposed above, energy and meteorology is an interdisciplinary area which offers exciting but complex challenges. Experts have been working and providing solutions for many years now (e.g. the ANEMOS' wind power forecasting system developed in the early 2000s by a European partnership). However, it is also clear that much more needs to be done. It is also becoming more evident that solutions, such as development of tools, require enhanced co-design and co-development between meteorologists and energy experts. In the context of climate services, the EU Copernicus Climate Change Service (C3S)[9] Programme is pioneering such an approach. Amongst the C3S projects, the European Climatic Energy Mixes[10] (ECEM) service is producing, in close collaboration with prospective users, a proof-of-concept climate service—or demonstrator. This C3S ECEM climate service demonstrator, comprising a set of tools, including an online web interface, for improved assessment of energy mix options over Europe, has been co-developed from scratch with extensive input from prospective users engaged through expert elicitation workshops and direct contacts. The main purpose of the C3S ECEM Demonstrator[11] is to enable the energy industry and policy makers to assess how well energy supply will meet demand in Europe over different time horizons, focusing on the role climate has on energy supply and demand. These are the types of activities that greatly enhance collaboration between energy and meteorology and at the same time produce valuable tools for decision making in the energy sector.

Other activities which foster collaboration, encourage critical thinking and promote innovative solutions are working groups within an organized environment. For instance, WEMC has recently launched its membership[12] which, as one of its core objectives, encourages meteorology and energy experts to engage in Special Interest Groups (SIGs).[13] The SIGs are vehicles leading to the production of reports, analyses and syntheses on key topics in energy and meteorology, which will ultimately assist the energy industry in addressing resilience, efficiency, mitigation and adaptation challenges.

Major Challenges to Be Addressed in a Co-design Approach

The Paris Agreement at COP21 has established a framework. Its impact depends on how the goals will be translated into real government policy actions. Nonetheless, the energy sector faces a critical need to transition from a carbon-based energy model to a decarbonized energy world. Low-carbon energy generation and energy efficiency are already key in this new paradigm. Hence, addressing the Energy Trilemma will require improved weather and climate information. A strong growth in renewables has already begun in electricity production and must continue; their contribution now needs to develop significantly in heat and transport (IEA 2016). As electricity is the easiest energy generation source to decarbonize, its share in the global energy production is expected to double in the next 50 years or so. The growth in solar and wind energy will continue. This will create new opportunities, for instance:

- The need for innovative solutions from start-ups and SMEs, in storage solutions and in smart management of energy systems at local to global scale;
- Job creation: the renewables sector is estimated to currently employ 8.1 million people (not including large-scale hydropower), plus an extra 1.3 million in large-scale hydropower (REN21 2016), and thousands of new jobs will be created by new projects;
- The development of distributed renewable energy will increase energy access, especially in the Asia-Pacific region and sub-Saharan Africa, where most of the 1.2 billion people who do not currently have access to electricity live.

But there will also be new or increased risks:

- The increasing share of wind, solar and hydropower is and will be reinforcing the dependence of energy systems on knowledge of climate variability and climate change. «Classical» energy systems need to become more flexible and to account for renewable variable generation (Cochrane et al. 2014). Hence, observational and forecasting capacities of the relevant variables need to be improved, on the necessary time and space scales. This includes local and high frequency wind and solar radiation forecasts, which are becoming more and more critical as the underlying variability poses problems to grid management, to ensure the real-time balance between consumption and production. Among the more specific challenges, one can (non-exhaustively) list:

  - Improved and more accurate characterization of past climate, especially in order to adequately assess wind, solar and hydropower resources. In addition to field campaigns during the due diligence stage of proposed projects, it is more than ever necessary to have multidecadal reanalysis at high spatial resolution, and at least hourly time step, to correctly assess the future performance of wind and solar farms. This allows better shaping of plant characteristics, and assessment of project bankability, ensuring a good return on investment, which is essential to make the sector profitable, and finally to improve its competitiveness with respect to other production means.

- Very short-term wind and solar radiation variations, on timescales of a few minutes to two to three hours. Ramp events, in particular, can destabilize power systems as they require a significant increase or decrease in production over short periods, to compensate renewables variability.

- Improved demand and generation forecasts on a few days to a few weeks, for optimized unit commitment planning and energy market operations. One key issue at the moment lies in the frequent jumpiness in successive weather forecasts, that is to say when consecutive forecasts give a different trend over the coming days. Reacting to these changes in real time often requires the buying/selling of energy in a sub-optimal way, leading to unnecessary expenditure.

- On longer timescales, from a few weeks to several months, improved sub-seasonal to seasonal forecasts would allow better planning of generation unit maintenance, and management of energy stocks, in particular hydropower capacity in large reservoirs.

- Of course, on longer timescales, energy companies and policy makers need improved information on the possible impacts of climate change on energy assets, and how future operations and systems management need to be adapted. This includes information on future means and extremes of different climate variables, together with the expected changes on the variability itself, as for instance a change in seasonality in precipitation will impact the yearly management of large reservoirs. Downscaling is of course a key issue because information is needed at a scale as close as possible to individual plants.

One could list many other examples of benefits arising from improving weather, water and climate information for the energy sector, and some of the other chapters of this book do so, building upon Troccoli et al. (2014). But, as mentioned in the previous section, improving weather and climate forecasts is not the only key to enabling more secure, more affordable and sustainable energy systems and services. Each component of the system needs to be taken into consideration in a global and integrated approach. For instance, the technical and economic analysis of the European electricity system with 60% Renewable Energy Sources study (Silva and Burtin 2015) shows the complexity of such an analysis, which requires many sources of information and different model components, with many interconnections between them. Improving the decision-making processes and the communication channels between energy and meteorology specialists is thus very important to ensure a coherent approach all along the chain. Therefore, a more effective integration of weather and climate information in energy systems requires:

- Improved communication between communities, at different levels (technical, managerial and decision/policy making). This should reduce the language gaps and enable a more rapid design of fit for purpose solutions.
- Common training programmes to inform energy people on weather and climate on the one side, but also, on the other side, for meteorologists to better understand how energy systems work, and how their inputs can be tailored to enhance operational models and decision chains.
- Closer and more responsive relationships between energy and meteorology people.

Among several other organizations, the World Business Council for Sustainable Development (WBCSD) emphasizes that pooling learning, exchanging best practice, sharing resources and encouraging mutual aid can benefit electric utilities and their stakeholders, as well as public authorities and consumers (WBCSD 2014). Increased sectoral and cross-sectoral collaboration is essential in moving forward and tackling the energy trilemma. Energy is now at the core of major programmes like the GFCS[14] and the C3S. Organizations with a strong interest in energy and in the role weather and climate have on it, such as WBCSD, the World Bank's Energy Sector Management Assistance Program (ESMAP), the International Renewable Energy Agency (IRENA) and WEMC, will play an important role in the future: to help develop science-based and user-driven solutions, for an effective integration of high-quality weather, climate and other environmental information into energy sector policy formulation, planning, risk management and operational activities; to better manage power systems on all timescales and strengthen climate change mitigation and adaptation.

NOTES

1. http://www.wmo.int/gfcs/.
2. http://www.euporias.eu/.
3. https://www.ametsoc.org/ams/.
4. http://www.emetsoc.org/.
5. https://windeurope.org/.
6. http://uvig.org/newsroom/.
7. http://www.wemcouncil.org/wp/conferences/.
8. http://www.wemcouncil.org/.
9. http://climate.copernicus.eu/.
10. http://ecem.climate.copernicus.eu/.
11. http://ecem.climate.copernicus.eu/demonstrator/.
12. http://www.wemcouncil.org/wp/about/membership/.
13. Typical SIGs could focus on (1) Weather/Climate Forecast/Projections for Energy Operation and Planning; (2) Grid Integration; (3) Data Exchange, Access and Standards; (4) Energy & Meteorology Education.
14. http://www.wmo.int/gfcs/.

REFERENCES

Brooks, M. S. (2013). Accelerating innovation in climate services: The 3 E's for climate service providers. *Bulletin of the American Meteorological Society, 94*(6), 807–819.

Buontempo, C., et al. (2014). *Climate services development principles*. ECOMS workshop findings. Retrieved December 14, 2016, from http://www.euporias.eu/sites/default/files/event/files/Climate%20Service%20Development%20Principles.pdf

Cochrane, J., Miller, M., Zinaman, O., Milligan, M., Arent, D., Palmintier, B., et al. (2014). Flexibility in the 21st century power systems. 21st century power partnership, NREL/TP-6A20-61721. Retrieved from http://www.nrel.gov/docs/fy14osti/61721.pdf

Dorling, S., & Bacon, J. (2017). *Developing a coupled weather, wave and tide forecasting system in support of construction and operations and maintenance of offshore wind farms*. 4th International Conference on Energy and Meteorology (ICEM 2017), Bari, Italy.

Dubus, L. (2014). Weather and climate and the power sector: Needs, recent developments and challenges. In A. Troccoli, L. Dubus, & S. E. Haupt (Ed.), *Weather matters for energy* (XVII, 528 p.). New York: Springer. ISBN:978-1-4614-9220-7.

Haupt, S.E., Kosovic, B., Jensen, T., Lazo, J., Lee, J., Jimenz, P., et al. (2017). Building the Sun4Cast system: Improvements in solar power forecasting. *Bulletin of the American Meteorological Society*. Early online release. https://doi.org/10.1175/BAMS-D-16-0221.1.

IEA. (2016). *World energy outlook 2016.* OECD/IEA. Retrieved from http://www.worldenergyoutlook.org/publications/weo-2016/

Mauser, W., Klepper, G., Rice, M., Schmalzbauer, B. S., Hackmann, H., Leemans, R., et al. (2013). Transdisciplinary global change research: The co-creation of knowledge for sustainability. *Current Opinion in Environmental Sustainability, 5*(3), 420–431.

REN21. (2016). *Renewables 2016 global status report.* Paris: REN21 Secretariat. ISBN:978-3-9818107-0-7.

Silva, V., & Burtin, A. (2015). Technical and economic analysis of the European system with 60% RES, EDF Technical Report. Retrieved from https://www.edf.fr/sites/default/files/Lot%203/CHERCHEURS/Portrait%20de%20chercheurs/summarystudyres.pdf

Troccoli, A., Boulahya, M. S., Dutton, J. A., Furlow, J., Gurney, R. J., & Harrison, M. (2010). Weather and climate risk management in the energy sector. *Bulletin of the American Meteorological Society, 6,* 785–788. https://doi.org/10.1175/2010Bams2849.1.

Troccoli, A., Dubus, L., & Haupt, S. E. (eds.). (2014). *Weather matters for energy.* New York: Springer Academic Publisher, 528 pp. Retrieved from http://www.springer.com/environment/global+change+-+climate+change/book/978-1-4614-9220-7

WBCSD. (2014, April). *Building a resilient power sector.* World Business Council for Sustainable Development Report.

WMO. (2011). *Climate knowledge for action: A global framework for climate services – Empowering the most vulnerable.* World Meteorological Organization, Report No. 1065. Retrieved from https://library.wmo.int/pmb_ged/wmo_1065_en.pdf

WMO. (2017). *Energy exemplar to the user interface platform of the global framework for climate services.* World Meteorological Organisation, 120 pp. Retrieved from https://library.wmo.int/opac/doc_num.php?explnum_id=3581

Open Access   This chapter is distributed under the terms of the Creative Commons Attribution 4.0 International License (http://creativecommons.org/licenses/by/4.0/), which permits use, duplication, adaptation, distribution and reproduction in any medium or format, as long as you give appropriate credit to the original author(s) and the source, a link is provided to the Creative Commons license and any changes made are indicated.

The images or other third party material in this chapter are included in the work's Creative Commons license, unless indicated otherwise in the credit line; if such material is not included in the work's Creative Commons license and the respective action is not permitted by statutory regulation, users will need to obtain permission from the license holder to duplicate, adapt or reproduce the material.

Index[1]

[1] Note: Page numbers followed by "n" refer to notes.

© The Author(s) 2018
A. Troccoli (ed.), *Weather & Climate Services for the Energy Industry*,
https://doi.org/10.1007/978-3-319-68418-5